THE ZAGHAWA FROM AN ECOLOGICAL PERSPECTIVE

THE ZAGHAWA FROM AN ECOLOGICAL PERSPECTIVE

Foodgathering, the pastoral system, tradition and development of the Zaghawa of the Sudan and the Chad

by

MARIE-JOSÉ TUBIANA
CNRS, Laboratoire Peiresc, Paris

JOSEPH TUBIANA
CNRS, Laboratoire Peiresc, Paris

A.A.BALKEMA/ROTTERDAM/1977

Exclusive Distributor:
ISBS, Inc.
P. O. Box 555
Forest Grove, OR 97116

*Dedicated to the memory
of Shartai Tijani al-Tayyib*

Translated from the French
by Philip O'Prey

© 1977 Marie-José Tubiana and Joseph Tubiana
ISBN 90 6191 015 3
Printed in the Netherlands

CONTENTS

	PREFACE	VII
	A NOTE ON THE TRANSCRIPTION SYSTEM	XI
1.	BIDEYAT AND ZAGHAWA TODAY	1
	The /bèRí/ in history	1
	Where are the /bèRí/ ?	2
	A consumers' economy	5
	A changing society	9
2	FOOD-GATHERING	13
	Wild cereals	14
	Wild fruits	18
	Other produce from gathering	25
3	THE PASTORAL SYSTEM AND THE NEED FOR TRANSHUMANCE	31
	The seasons	32
	Vegetation	34
	The water supplies	37
	Livestock	39
	Rights of ownership and use	45
	Seasonal movements of flocks and herds	49
	Herdsmen and their way of life	72
	Animal produce	78
	Conclusion	81
4	TRADITION AND DEVELOPMENT	83
	1. Improving the pastures	84
	2. The possibilities for modifying agriculture	85
	3. Introducing arboriculture	86
	4. Improving horticulture	87
	5. The water problem	87

6.	Craftsmanship and small scale industry	91
7.	Education	92
8.	Conclusion	93

Appendix: APPETENCY INDEX 95

NOTES 99

SELECTED BIBLIOGRAPHY 107

INDEX 111

PREFACE

It was in 1956 that we first came into contact with Zaghawa society in the Eastern Chad. Our only guidance at the time came from Sir Harold MacMichael's studies on the Zaghawa of the Dar-Fur and Captain Chalmel's article on the Bideyat of the Ennedi. Thanks to the goodwill of the majority of the French administrators and to cooperation from Zaghawa and Bideyat of every rank and position, the thirteen months or so of fieldwork were to prove an extremely fruitful period. It was nevertheless quite some time before we had the opportunity, in the course of three stays in 1965, 1968-69 and 1970 respectively, of discovering for ourselves the Zaghawa of the Sudan. The welcome we then received in the Sudan, both from our colleagues at the University and the Zaghawa themselves, enabled us to work in the very best conditions.[1]

In undertaking the study of Zaghawa society, we set ourselves the goal of achieving a comprehensive description of this society, situated in a historical perspective.[2]

Actually the studies we have so far published only represent a few sections of the entire project. The fact is that the publications have appeared in an order that was anything but planned, under the influence of various circumstances. Normally we would have begun by publishing either the present work or something similar.

As anthropologists we believe that the first step in this sort of scientific investigation must surely consist in studying the ways and means used by a specific society in order to enable itself — that is to say, its members — to achieve existence. This point is often disregarded, although the basic assumption that one cannot study a society which has not achieved existence must of course be considered a trivial statement. It is self-evident that men cannot afford to build a culture before they have fed themselves. Three factors must be taken into account when studying the living conditions of human groups: a) the natural environment, the resources it provides and the difficulties it presents; b) the level of technical achievement, that is to say the capacity of man to exploit the resources offered by

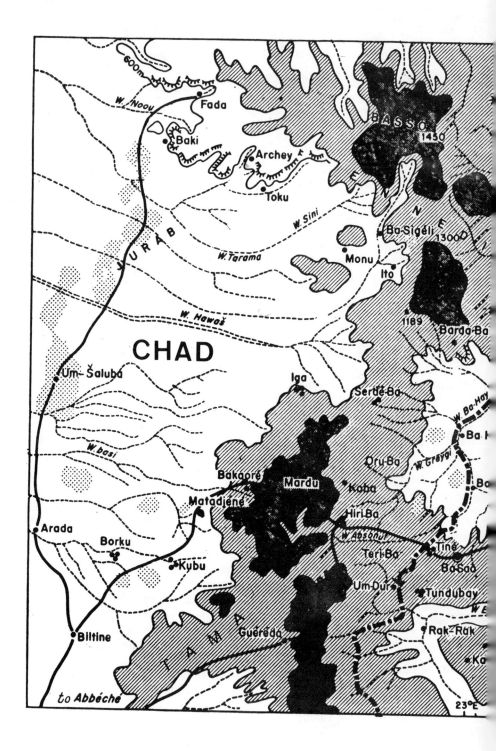

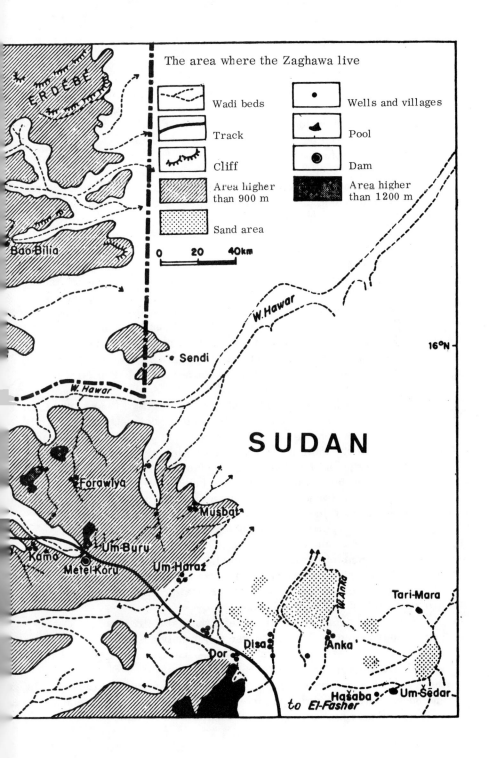

the natural environment, with its decisive impact on production;
c) lastly, the economic system, which controls the production of
goods and their distribution, and consequently possession, exchange
and consumption. As our reader will understand, we cannot in this
short preface develop this argument. Suffice it to say that the general
scientific approach which underlies the various aspects of our
research is inspired by the teaching of Marcel Mauss.

In this short volume then, we have collected four articles, appropriately altered or shortened, that have appeared separately in French. In this language they were accessible only to the Zaghawa of the Chad, not to those of the Sudan. It is for them and also to fulfil a wish, which both they and our friends and colleagues of Khartoum frequently expressed, that we are publishing these articles now in an English translation.

The first chapter was written by us with just such a book in mind, so as to present today's Zaghawa society in its modern context. Marie-José Tubiana wrote the next two chapters. Chapter four, the initiative for which is in some sense due to the Zaghawa themselves, was jointly written for a conference on pastoral societies that took place at Niamey in December 1972. We found that the scientific study of the Zaghawa society has led us to be more and more concerned with their present distress and anxiety, and ready to co-operate in the research for appropriate measures to palliate them, when we were invited to do so by our local friends.

A fifth article has not been translated into English. It appeared in the "Cahiers d'Etudes Africaines" in 1962 and it describes a state of affairs which has by now considerably altered. Starting with a study of the markets, it laid out the main lines of the economic interdependence of the populations living on the Chad-Sudanese border. It would have been necessary to bring this study up to date to include it in this book.

To mention everyone who has helped us would make a long list and there would almost inevitably be omissions. It gives us great pleasure to take this opportunity of thanking them all. In order to mark our gratitude we have felt it appropriate to dedicate the present book to the memory of a man of considerable intelligence and drive, who was to be a steadfast friend throughout our stays. We would also like to commemorate on this occasion Ibrahim 'Fransawi', another Zaghawa who proved a faithful and selfless friend.

We wish to record here our grateful thanks to Professor I. Cunnison and our friend F. Rehfisch, who generously agreed to read the manuscript and gave us the benefit of their advice. Any errors that remain are entirely our own responsibility. Lastly we would like to thank P. O'Prey, our translator, as well as the editors of the journals and book where the original articles were published, and Mme M. Lefèvre, who so patiently typed the English manuscript.

A NOTE ON THE TRANSCRIPTION SYSTEM

a) Zaghawa (Z.)

/bèrí-a/ or "the language of the /bèrí/" (Zaghawa and Bideyat) is a tone language. However tones are not marked unless we are certain or nearly certain of our transcription.

High tone is marked with /´/ e.g. /á/. Low tone is marked with /`/ e.g. /à/. Rising tone is marked with /ᵛ/ e.g. /ǎ/. Falling tone is marked with /ᴧ/ e.g. /ã/.

Vowels — Vowel sounds are five in number: /i/, /e/, /a/, /o/, /u/.
/i/, approximately like the "i" in the French "si", or in the Arabic 'i' of medium length.
/u/, approximately like the "ou" in the French "fou" or in the Arabic 'u' of medium length.
/a/, approximately like the "a" in the English "father", or in the French "papa" or in the Arabic 'a' of medium length.
/e/ and /o/ are not very closed; they resemble the Sudan Arabic 'ē' and 'ō', but are of medium length.
There are no nasal vowels. The vowels nasalised by the proximity of a nasal consonant may be superscribed by ˜

Consonants — The signs have the same value as in English. But:
/g/ is always pronounced as in the English "gate";
/j/ denotes a soft consonant situated between the English "j" and "dy";
/h/ is always aspirate;
/ň/ corresponds to the French "gn" or the group "ny" in the Sudan orthography of place-names (e.g. "Nyala" would have been written /ňala/);
/ŋ/ corresponds to the "ng" group in the English "meeting";
/ʀ/ is a retroflex (flapped) "r";
/š/ corresponds to the English "sh".

b) Arabic (Ar.)

The Arabic words are recorded in the local pronunciation. They are transcribed between single quotation marks according to the same principles as the Zaghawa words. In addition the following are used:

'x' to denote the breathed postpalatal plosive corresponding to the Spanish "jota" (Arabic خ)

'h' to denote the breathed glottal (Arabic ح)

Long vowels are marked in the usual way: 'ā', etc.

c) Place names

Place names not given in transcription follow the official maps' orthography. When transcribed, they do not take a capital letter.

Chapter 1

BIDEYAT AND ZAGHAWA TODAY

The Arabs give the name Zaghawa to an African population which calls itself /bèRí/ and lives on the border of the Chad and Sudan Republics, around latitude 15^0 N and longitude $21^0 - 25^0$ E. The same Arabs use the name Bideyat for another African people that live around the $17^0 - 18^0$ N who similarly call themselves /bèRí/. This is no coincidence since, besides a common language and culture, close economic ties together with many other features characterize both groups which in fact form one and the same population.

THE /bèRí/ IN HISTORY

The earliest mention in an Arab source (Ibn Qutaiba, quoting Wahb ibn Munabbih, in his "Kitāb al-Ma'arif") of the presence of the Zaghawa in the Eastern Sudan goes back to the 8th century A.D. However the past of the /bèRí/ remains fairly obscure, no doubt because modern historians have followed very much the same line as their Arab predecessors, scrutinizing their remote origins rather than the recent past.

In fact we do not know if the present day /bèRí/ have the slightest thing in common, apart from the name, with the Zaghawa of the Arab writers.

They interest us, as anthropologists, in so far as we can observe them and describe what kind of people they are, the country where they live and how they manage to live there.

But we intend to take into account the fact that these people have had a history of their own, consisting of a series of climatic events, of a gradually acquired technical knowledge, of the development of relations among their own clans and also of their relations with the neighbouring African populations.

Furthermore we know that over the years the /bèRí/ have had to endure outside pressures of some importance, all of which have eventually reinforced the processes of islamisation and arabicisation.

Begun approximately around the 16th century A.D. with the arrival of Muslim preachers and of small groups of Arabs or arabicised Africans in the neighbourhood, these processes were later strengthened and complicated by attempts at political domination in the area by the sultanates of the Dar-Fur in the east and Wadday in the west. With the British and French colonial conquests in the 20th century, that have readily maintained the rivalry of the subjugated sultanates, the division of the /bèʀí/ people is written into the international treaties and has become (at least for a time?) definitive. Time was too short to allow the Libyan brotherhood of the Sanusi to establish themselves among the /bèʀí/, but the colonizers have reinforced and speeded up the islamisation and arabicisation of the /bèʀí/, both deliberately and unintentionally.

Under the independent states which took over from the colonial powers the policy with respect to the /bèʀí/ people has hardly altered. The latter, who already knew how to exploit to their own advantage the artificial division created by the border, soon learned to play their own part in the political systems set up after independence. All complained of being neglected and under-equipped by the central government, compared with the rest of the country to which they belong; nevertheless everyone considers that the government shows far too much interest in them, especially where taxes are involved. There can be no doubt that the existence of this discontent has led the former Chad government to suspect the /bèʀí/ strongly of having a certain sympathy for the armed revolt of the National Liberation Front of the Chad (FROLINAT).

WHERE ARE THE /bèʀí/?

The natural environment in which the /bèʀí/ live is harsh and oppressive. Its constraints may be read in the landscape, and are accounted for by the climate. Across impoverished soils overhung by a cloudless sky from which very little rain ever falls, wander cows, goats, sheep and camels in herds or flocks. In the north, east and west, the desert hems them in. The south is scarcely more hospitable, occupied as it is by settled farmers, and the area beyond is infested with the tsetse fly.

South of the geographical limit of the Sahara are situated the Ennedi highlands which form the heart of Bideyat country. They consist of a group of plateaux separated by deep gullies. The altitude, which is never less than 600 m, reaches 1,450 m in the Basso. There a comparatively large amount of rain accounts for the presence of a fairly dense vegetation consisting of grassy stretches and fine trees, where animal life abounds. Over a number of hours, even days, wadis run from this meagre reservoir of ochre sandstone to be absorbed in the north by the depression of the Mourdi, in the west by the outer reaches of the small Jurab basin, and in the east by the valley of the wadi Howar. Within the highlands underground streams supply springs and permanent pools.

South of the Ennedi the crystalline shelf comes up to the surface. This is the Zaghawa plateau, which exceeds 600 m, averaging an altitude slightly below that of the Ennedi (between 800 and 1,000 m) and nowhere reaching the height of the Basso. It constitutes a northern extension of the crystalline range of Wadday. As it joins on to the Ennedi, it is at first hemmed in between the divergent channels of the wadis Howar, flowing east, and Hawash, running in a westerly direction; then, as one goes southwards, it widens considerably. Here the rains are more abundant but just as irregular. The water seeps quickly into the sandy soil. The clayey beds form temporary or permanent pools. The waters flow eastwards, towards the bed of the wadi Howar (known successively as Tiné, Ba-mina, and Ba-hay) and westwards towards the plain which continues the Mortcha to the south, where Mahamid and Daza live a nomadic life. The vegetation is not so rich, the animal life not so evident, but the pastures are more extensive and the cultivation of small garden-like plots in the dried-out beds is less risky. The country is more densely populated.

The southern boundary also constitutes a climatic one: the area where agriculture is possible is firmly held by the Mimi, Mararit, Tama and Gimir peasants. The Zaghawa only encroach onto the small district of the Guruf on the far south-west corner of their habitat, where they come down to an altitude below 600 m.

More to the east, in the near right-angle formed by the bed of the wadi Tiné (later Howar) the shelf rises above 900 m, and the valleys never go below 600 m. The pattern of the rains is identical: these get scarcer as one goes eastwards; the divergences with western Zaghawa country are due rather to human factors: the extermination of the fauna, the extension of cultivation (however risky), and the construction of a few artificial pools. The waters flow westwards and northwards, towards the wadi Howar.

On the southeastern border of Zaghawa country is the volcanic complex of the Jebel Marra, the territory of the Fur peasants, with its rocky inselbergs that break the monotony of the landscape. In actual fact this varies a good deal according to the types of soils and their ability to retain moisture throughout the nine dry months of the year. The normal aspect is one of vast sandy stretches from which rise occasional picturesque bare sandstone outlines or small rusty-coloured granite hills broken up by erosion into boulders equally bare, and casually piled on top of one another. These might be pastures of dry standing fodder, fields of wild grasses with thorn trees covered in foliage, occasionally millet fields that have been harvested or have failed to ripen. The beds of the clay basins look peeled, without a single blade of grass, baked in the sun like bricks and with the tracks of animal hooves imprinted in them. The wadi beds spread out beneath huge shady trees with their wells, drinking troughs and occasional irrigated gardens. The horizon is never bare, whether there rises from it a long ridge of angular rocks like the bent spine of some prehistoric monster, or a sharp peak pointing skywards like a finger. In other places it may be restricted by the quite sharply outlined ledge of a higher plateau, giving from afar the impression of a young mountain range, as in the Kabka.

Each year sees the return of the rainy season, /gyé/, which begins roughly end of June and finishes late August. The first showers may begin in May; the last ones never fall later than September. The temperature is chilly and the colours change quite rapidly. The sky becomes grey and overcast with clouds. Wadis flow, pools form. The ground becomes covered with greenery. Certain of the trees shed their leaves. Grass grows knee-high. Yet it rains more in the south than the north, and more on the heights than on the low areas.

In the Sudan, Zaghawa country has an annual rainfall of between 300 and 150 mm; in the Chad between 400 and 150 mm, while the Ennedi has between 150 and 100 mm.[3] When the rains are inadequate or absent, the /bèʀí/ retreats southwards in search of pastures. When he is reduced to seeking shelter with the farmers, he takes his herd or flock with him, but it will not normally survive and he will take a job as an agricultural labourer in order to subsist. During the great famine of 1969 and again in 1972-73, the Zaghawa of the Sudan went to settle as far as Hofrat en Nahas. When abundant rains have covered the "sahara" with fresh pastures, the /bèʀí/ leave their country by the north and east to take advantage of them, and then return to the centres from which they set out. In this way the /bèʀí/ habitat contracts and expands following the weather or, as R. Capot-Rey puts it, "water's tyranny".

Next comes /tằrbà/, the season of harvests, a short, dry and hot period, where yellow gradually replaces the green. The harvesting, which will go on until December, begins in October. The deciduous trees recover their leaves, there is water in the wells, pools begin to go down and animals are fat.

Then the temperature gets colder with icy nights and a harsh wind blowing from the northeast. The sky is overcast. On the ground the yellow of the dried grass predominates. This is the dry and cold season, /dằbó/, which lasts from November until January.

1. Cattle watered at Anka wells. October 1965.
2. Wadi Absonut running full after a violent downpour. August 1957.

3. Dor. Flocks of sheep returning from the wells. February 1969.
4. Um Shedar. Cattle drinking from the pool. December 1965.
5. Hiʀi-ba. Women threshing boù in kobé country. Notice the wooden flails, the small brush and winnow. October 1956.

6. Hiri-ba. Winnowing wild cereals. October 1956.
7. Hiri-ba. Market. 1957.
8. Dirong. Outside granaries. 1957.

9. Drawing water from Koba well. February 1957.
10. Metel-Koru dam (south of Um Buru). January 1969.

11. Bakaoré. The village. — 12. Inside of a hut. Grindstone and cloth placed under the woman's knees when she grinds. In the background, storage jars.

13. Anka. Woman carrying water in a dipper to fill her trough.
14. Marugwi (a ferik). Making butter in a calabash. February 1957.
15. Millet field. October 1950.

February sees the beginning of the hard ordeal of /áîgî/, the dry and hot season, which, God willing, will end in June. This is a critical period of gradual impoverishment: the dryness of the country is unmistakeable. The first rains are anxiously awaited. The traces of those of the preceding /gyé/ have disappeared. If Allah grants a few showers in May, it will help bridge the gap by giving a start to the vegetation. Otherwise it will be June or sometimes July, before that happens.

Thus the /bèʀí/ have to pass through at least nine months of dry season before reaching a short rainy season of such irregularity that several entirely dry years may follow in succession. In some years the total rainfall is inadequate; in others, while the total may be quite high, the falls of rain are insufficiently spaced apart to allow the renewal of the pastures.

When the rains are abundant and wide-spread, the upper limit of the pastures shifts considerably northwards and /bèʀí/ country annexes a vast part of the "sahara" desert. When the rains are poor or follow too closely, the usable area shrinks accordingly. Flocks and herds then go south, immediately raising the awkward problem of co-existence with the settled population living there, for the latter, just like the nomads, have animals to water and feed and their fields are not spared by the drought.

A CONSUMERS' ECONOMY

The meagre resources of the natural environment closely affect the living conditions of the /bèʀí/. Their economy may be described as one of consumers, in so far as they are non-producers. Their main effort indeed is put into gathering natural products, not into an organised exploitation of their environment. The /bèʀí/ harvest the ears of wild grasses, pick wild fruits, unearth tubers and more recently have started collecting the gum of "Acacia Senegal"[4]. The latter is put on sale, whereas the former are eaten.

The /bèʀí/ especially the blacksmiths, used to hunt in order to subsist. The tanned hides were used or sold. However, in the Sudan, the wide distribution of fire-arms has led to the almost total extermination of game. The precise situation in the Chad is not quite clear.

Animal husbandry may be regarded as an attempt to organise natural processes: the herds and flocks are guarded against thieves and wild animals, watered (a considerable task in the dry season), taken to pastures (wherever grass and water may be), cared for, and periodically vaccinated (ever since the British and French governments took an interest). There is no need to store up hay: the dry climate prevents the standing fodder in the steppe from spoiling. The camels can browse off the trees on their own; the goats occasionally try to climb them, though the men knock down leafy branches and green pods for them and other less agile animals.[5] Those beasts likely to be attacked by wild animals are brought together for the night in rough and ready pens consisting of an impenetrable enclosure of thorns.

Milk and butter are consumed, meat rarely so; sheep and cattle are sold for slaughter and camels as remounts when prices are high. The predominant interest that the Zaghawa have in their livestock is as economic assets; there are no indications of the kind of relationship associated with the East African "cattle complex".

More bulrush millet fields are to be seen in the Sudan than the Chad. In the Sudan they are often large community fields, made up of individual plots but small individual fields may be found in both countries. However this type of agriculture, which makes no great demands, remains little known. European observers have often blamed the /bèRí/ for their idleness, when they ought rather to have emphasized their strong awareness of economic realities. For in this region where even drought is nomadic, the cultivation of bulrush millet is a gamble verging on the absurd. The statistical probability of reaping as much as was originally sown is almost nil. The loss felt hardest of all will not be that of the few days of work required for preparing and fencing in the ground, sowing, weeding and keeping watch on it, but that of the seed bought and seldom recovered. The attitude of the /bèRí/ with regard to agriculture is thus perfectly sensible.

Those who sow tomatoes, onions, pimentos or okra in a wadi bed after the rains, know they will obtain a crop, if they only bother to dig a well nearby and water daily this garden. The /bèRí/ does not begrudge his labour when he can expect a result.

Those who are concerned to encourage agriculture among the /bèRí/ would be better advised to urge them to sow seeds of the less exacting wild grasses on suitable soils: failure is unlikely and any eventual loss minimal.

Where possible, the /bèRí/ exploit the subsoil, as they have done for a considerable time. In this way visits are regularly paid to natron and rock-salt mines, whose product is either consumed or exported.

The blacksmiths have stopped collecting iron ore, since scrap metal has become plentiful.

Naturally the /bèRí/ use the underground water through a large number of wells, some of which are timbered on account of their great depth, constituting remarkable specimens of engineering.

Skilled craftsmanship is in the hands of a despised caste, whose men are blacksmiths and whose women are potters. They make the metal parts of the few agricultural tools used (digging and weeding hoes, heads for digging sticks, axes, and harvesting knives for women), or weapons (daggers, spears and swords) or jewelry made from copper, aluminium or silver as well as their own tools for the forge. They burn their own charcoal. In the Sudan the /bèRí/ blacksmith acts also as a tinsmith: empty tins are turned into water jugs, coffee jugs and even small chests. He also does woodwork: pestles and mortars, stools, pack-saddles, the bodies of musical instruments, etc. He can tan hides and work leather; out of gazelle sinews he can make nets which he uses out hunting with his dogs, thus making him a supplier of fresh or smoked venison. In the past he used to weave the wild cotton that was gathered and could construct his own loom.

The wives of the blacksmiths make all the fired pottery.

With regard to services, the blacksmith plays the drum and his wife is the women's hairdresser.

All the /bèRí/ men can make wooden objects and vegetable rope, and spin cotton and sew. All the women can make unfired pottery and sewn basketwork. They co-operate in building thatched houses of stone and mud.

The demand for manpower in /bèRí/ country is non-existent, except for the Sudan where a few herdsmen, lorry drivers, mill workers, etc. are employed. The man who wants to hire out his labour goes out as a day labourer to the neighbouring farmers or far away to the cotton plantations of the Gezira Scheme, or even as a navvy to the sites of public building projects.

A small class of modern capitalist businessmen is in the process of being formed, especially among the Zaghawa of the Sudan. Its members come from the traditional chiefly families, who are better able to appreciate modern developments and obtain grants. Their enterprises are quite modest: trading, transport, milling and building.

As regards food, the /bèRí/ import bulrush millet or sorghum for which there is a constant need, sometimes in quite considerable quantities, from the neighbouring farmers. Sugar and tea from abroad are luxury products, as are dried dates coming from palm groves that can be close by or quite distant.

All these products may be found in the markets that take place weekly in the Chad at Hiri-ba, Bakaoré, Matadjéné and since 1959, Oru-ba and Serdé-ba, and in the Sudan mainly at Disa, Dor, Um Marahik, Musbat, Um Buru, Kornoy and Tiné. Zaghawa country thus appears to be a meeting-place for cattle-breeders from the north and farmers from the south.[6]

As for clothing they now hardly ever use handwoven cotton but rather imported materials, cotton and rayon. Other manufactured products from outside are cotton blankets, towels, plastic sandals, matches, cigarettes, teapots, trays, enamelled metal or aluminium pots and basins, glasses and cups, dry batteries, aspirin, soap, etc.

Exports take different routes: their settled neighbours take the salt, natron and animals on the hoof, whereas flocks of sheep or herds of camels are sent far away to the large urban centres, even as far as Omdurman. The animals often reach Egypt or Libya.

The advantages of border trade have not escaped the /bèʀí/ who know how to cross the border to sell at the highest price and buy at the lowest. They rarely bother with the superfluous formalities of the customs post.

The theft of camels from the neighbouring peoples, which may be regarded as a form of free import, is increasingly rare in practice.

In a country with such long distances, trading means transporting. When the main problem used to be that of security, the /bèʀí/ hardly went beyond the boundaries of their habitat and the exchanges with their neighbours were governed by established custom. The transporters used camels and donkeys according to the loads and distances involved. Nowadays some of the /bèʀí/ have become traders. They buy camels and sheep in order to sell them and import bulrush millet, manufactured goods, tea, etc. The lorries, which they occasionally own, travel about a country which is almost entirely lacking in roads and has just a few tracks.

Barter, which until a few years ago continued to be favoured by the /bèʀí/, is on the decline in the Chad and is almost non-existent in the Sudan where all transactions are carried out in the local currency.[7]

Money is injected into the /bèʀí/ economy in a number of ways. On the one hand, /bèʀí/ and foreign traders purchase local products in order to sell them again, and on the other the public officials (police, nurses, veterinary surgeons, teachers, etc.), some of whom happen to be /bèʀí/, receive a salary which they spend locally on their various personal needs. Furthermore the traditional chiefs receive a salary of which a part goes into the local circuit when they purchase imported goods or offer gifts to their wives, children and clients.

This money is spent through taxes or purchases in the market, and on journeys, for people enjoy travelling by lorry.

A CHANGING SOCIETY

While the technical level of agriculture and craftsmanship is not very high, the activities of gathering and husbandry, including all those things dependent on them, pre-suppose a sophisticated knowledge of the natural environment, covering botany (together with some knowledge of soils), animal husbandry (including the reading of animal tracks), meteorology, routes, watering points, etc. Men and animals adapt themselves as closely as possible to the environment. The flocks and herds require grass and water in order to subsist, but not all the species possess the same endurance and they cannot all travel at the same speed. Furthermore, when the grass is green, the animals drink less than when they graze on dry straw. In order to guide and protect the flocks and herds as well as water and milk them, the /bèRí/ have to accompany them during their transhumance. As a result /bèRí/ society has a two-fold way of life (or "double morphologie" as Marcel Mauss used to say of the Eskimos), whereby permanent villages and temporary camps are occupied by turns.

The first rains see the /bèRí/ gathered together in the neighbourhood of the permanent wells, under the shelter of their round straw-thatched stone and mud houses. Here also are to be found the granaries, now empty or almost so. The youths will leave with the camels and sheep for the traditional transhumance routes as soon as the re-appearance of the pastures has been confirmed. They may be absent for several months at a time, living mainly off milk. The cows and goats remain in the proximity of the village and are milked daily. At the beginning of the dry season, those left in the villages camp out with cows and goats and disperse under straw huts among the dry wadis to take advantage of the pastures and the permanent wells. Such is the appearance of /bèRí/ society during the times of plenty. Then comes the end of the dry season together with privation. People gather close to the permanent wells and re-occupy the villages, awaiting the rains, while they share frugally the remaining resources.

One can regard the /bèRí/ village as having been at first merely a spatial projection of the political and religious unit that is the clan. All the members of the clan lived there under the authority of a chief descending from the ancestral founder, himself personally present to protect his descendants in a sacred tree, rock or mountain, the site of the seasonal or enthroning rituals. These clans were not isolated, for exogamous marriage, which was the rule mainly among the Zaghawa, would set up alliances between them. Those taking wives would pool their resources to provide a bridewealth as high as 100 cows which the members of the bride's family would share out among themselves. These gifts might be spread over several years and were reciprocated by symbolic counter-gifts. The chiefs' authority could extend beyond their own clan to smaller or poorer clans.

However the whole of this picture, which can be glimpsed in the actual situation today, has been considerably altered by the impact of Islam, followed by that of European colonization. The adoption of Islam by the /bèʀí/ took place gradually, occurring somewhat earlier among the Zaghawa of the Sudan than among those of the Chad, and more recently among the Bideyat. By virtue of its universal and monotheistic character, the Islamic religion breaks down clan structures, for the villages accept strangers into the clan more easily. The worship of ancestors is given up (for instance the sacrifice of pregnant animals) or else re-interpreted so as to be acceptable to the orthodoxy represented by learned men ('faki'); as a consequence people are less reluctant to leave the village. Nevertheless the prohibitions continue on the whole to be heeded and exogamous marriage continues in practice, although in theory it has been discarded in favour of the preferential form of marriage to father's brother's daughter. Polygyny is restricted to four wives, but Islam allows concubines. In fact there are few polygamous /bèʀí/ with the exception of the chiefs. Circumcision, a traditional practice, is confirmed by Islam. When disputes are brought before the official courts, justice is dispensed according to the Maliki system, but custom has retained its rights. Recourse to vendetta remains current practice. Islam has furthermore introduced the art of writing; it has also raised the status of study and commerce, the latter a respectable activity reconcilable with the deepest piety. It has given rise to a small elite of keen jurists.

With the support of the French authorities the idea of a hierarchy headed by a supreme chief levying taxes has been accepted much more readily in the Chad than in the Sudan. Whereas in the latter the Zaghawa are distributed among seven chiefdoms (three of which are major ones), in the Chad on the other hand there is only one sultan, chief of the Kobé Zaghawa. He has under his authority the three other chiefdoms, at whose head he tries to place his own henchmen. Among the Bideyat the colonial power strove to strengthen the authority of two main chiefs whom it required to be submissive.

Confronted with all these changes, women's behaviour remains more conservative than the men's. Women continue to frequent the places of worship, heed the prohibitions and seek the aid of magic.

The development which we have attempted to describe in a schematic fashion is far from over. Its effects are more evident among the Zaghawa of the Sudan than those of the Chad, the Bideyat being the least affected. Here we have simultaneously before us three successive stages of the same developing process. For there is nothing fixed in /bèʀí/ society, which has assimilated the idea of progress from the colonizers and intends to go on improving its living conditions with remarkable energy.

Original French article, 1973: "Un peuple noir aux Confins du Tchad et du Soudan: les Beri aujourd'hui", Les Cahiers d'Outre-Mer, 26(103):250-261.

Chapter 2

FOOD-GATHERING

Still the /bèʀí/ have not waited for the arrival of the colonizers to borrow the practice of agriculture from their southern neighbours. This remarkable development is mainly illustrated by the cultivation of millet (/bàgà/, Ar. 'duxn': bulrush millet), more widely spread among the Zaghawa of the Sudan than those of the Chad, where it is now a feature of the Guruf district; the Bideyat have been unable to adopt the practice for the good reason that the conditions of their country could not allow it.

Prior to the introduction of millet agriculture the /bèʀí/ used to meet their needs in cereals by entrusting their womenfolk with the task of gathering wild grains. Nowadays however it seems that many of the men feel that millet cultivation should definitely put an end to the gathering of wild foods. We had no opportunity to inquire into women's attitude on this question, but their behaviour provides a sufficient answer in itself.

In the Chad, where the Kobé are prominent, the gathering of cereals, berries and wild fruits still forms an important part of female duties and of the activities of young herdsmen. For women, it is a way of providing a supplement, an indispensable one at that, to the family meal. The produce gathered, in particular the cereals, is generally stored in granaries and then turned into flour and porridge. The food gathered by the herdsmen consists mainly of berries and other fruits that are eaten raw on the spot. In this way they vary and improve their daily diet by means of a genuine nutritive supplement.

In the Sudan, where millet grown over large areas usually covers the demand for cereal food and where there is a greater volume of trade, the gathering of food by women has now fallen into disuse, except in the extreme western part of the country, where some of the Kobé live. Nevertheless the herdsmen and young Sudanese children continue, as before, to gather wild fruits. Yet during the recent years of drought and famine, when agriculture totally failed, the gathering of anything that is edible was quite naturally resumed.[8]

This dissimilarity between the Sudan and the Chad /bèʀí/ accounts for the fact that the research material used here comes for the most part from the Chad.[9]

WILD CEREALS

/bôû/

Of all the grasses that make up the vegetable cover of the Zaghawa region, /bôû/ (Ar. 'absabe'; Dactyloctenium aegyptiacum (L.) Beauv.: Gillet, No. 371) is certainly the most providential. It grows nearly all over the wadis and the sandy soils, especially in the south; there, at the end of August, the women begin their harvest during that difficult transitional period, when one must live on one's last stocks, waiting for the first ears of millet to ripen. They may continue the harvest over two months, or more in a good year. In general they harvest the /bôû/ twice with an interval of a fortnight between the two sessions; but the second harvest is smaller than the first.

The gathering of /bôû/, just as with all wild plants, is a task for women. Young or married women set off alone or in small groups for the spot judged to be rich in this plant and settle there for a fairly long period in order to bring back as many as possible of the tiny dark brown grains in the form of a crop which may occasionally amount to three or four camel loads (approximately 130 kg per load).

The gatherer mentally delineates a certain area covered with /bôû/, whose stalks she cuts off with a stick if they are quite dry, or else with a short knife. She then gathers her harvest by means of a small broom: /jórdò/, made from a bundle of dry stalks. Next she separates the ears from the straw by taking a handful of /bôû/ stalks between her hands and the broom and letting them fall in the wind. Other women pull off the /bôû/ stalks using a rake with a very short handle: /ókkò/ which they make themselves from branches tightly bound with pieces of bark.

The heaps of ears are for a time left on the spot and lie at intervals behind the woman who continues her reaping. To keep animals away, she surrounds the heaps with thorny branches and to protect them against theft she may lay on the top a /mándà/ stone, a symbol of the sacred mountain abode of the "ancestral spirit" protector of her clan,[10] or she may perform an act of tutelary magic.

The gatherer's work may last a month, sometimes more. The women, who have left home with all that was necessary for their subsistence, spend the night under rudimentary shelters, which they build from branches and bundles of grass. From time to time, they pay a visit to the nearest market in the hope of hearing any news or to purchase some supplies.

When the gathering is over, the ears can be taken back home or be threshed on the spot, as seems most often the case. With the aid of a flail /dùgún/ which she has brought, the woman threshes the ears with the flattened side of the implement's head. The same flail is used to thresh millet. According to one woman, they sometimes separate the grain simply by crushing the ears in the hollow of their hand. The grain is then winnowed and put in leather skins. A man belonging to the family, generally a son, will come and fetch them with a donkey or a camel according to the quantity harvested. The /bóù/ grain is kept in huge earthen storage jars, located within the dwelling, separate from the cultivated millet: "for one must not mix cereals within the same granary".

/bóù/, like other wild cereals gathered, must be regarded as a millet substitute. Clearly if the millet harvest sufficed or if there were good possibilities of obtaining enough, women would spare themselves the trouble of going off to harvest the /bóù/, as is shown by the investigation carried out among the Zaghawa of the Sudan.

The porridge made from /bóù/ is not nearly so delicious as that made from millet, even though it is slightly sweet. Its nutritive value is perhaps inferior. In any case, when a woman has only /bóù/ and very little millet left in her storage jars, she uses the millet to prepare the food for her husband and sons-in-law, keeping the /bóù/ for herself and her children.

Besides porridge, women utilise /bóù/ to make a kind of dough: /diŋe/ with which they rub and clean their bodies. Cooked and dried, the flour made from /bóù/ is mixed with perfumes, then moistened at the time of use to make a ball of dough.

The different kinds of 'kreb'

The Arabic speaking people of the Chad use the name 'kreb' to designate different grasses which the Zaghawa call /áìrî/, /jîgîʀì/ and /sábà/.

The first two are equally known under the more general name of /gú/;[11] they are harvested twice at a month's interval like /bóù/. They could well be identical with the Echinochloa colona Link., recorded by Gillet in his catalogue (No. 375) under the Bideyat names of "airi" and "digiere". The grains of the /áìrî/ and /jîgîʀì/ are tiny; they are harvested in different ways. As far as /áìrî/ is concerned, the plant is beaten with a flattened branch before it has dried and the grains fall into a basket in which they are collected. With /jîgîʀì/ the grasses are cut with a reaping-hook, the harvest is gathered with the aid of a broom and the grains are separated from the stalk exactly as with /bóù/. The grain is pounded in a mortar, then crushed to flour just like /bóù/ grain.

The wild grass most appreciated by the Zaghawa: /sábà/ is without the slightest doubt Panicum laetum Kunth. (Gillet, No. 388). It is harvested just once, by cutting the stalks with a reaping-hook.

We have noted the name of a fourth grass: /ègé/, for which we were again given the Arabic equivalent of 'kreb'.

They are all prepared like millet in the form of a porridge eaten, whenever possible, with a sauce made from meat, dried tomatoes, onions, okra (Hibiscus esculentus), and red peppers, or even served with milk.

Other wild grains

/búbù/ (Ar. 'am-hoy' in the Wadday; 'kwoinkwoin' in Dar-Fur; Eragrostis pilosa (L.) Beauv.; Gillet, No. 383) is a plant found in the neighbourhood of wadis which is harvested over a very short period, about a week after the rains. Its grains are very small; they are gathered like those of /áìrì/ by being knocked into a basket with a flattened branch. Then they are ground in a mortar, winnowed and crushed on the grindstone. A porridge which is rather popular is made immediately afterwards from this flour. /búbù/ grain is not stored in granaries.[12]

Another (unidentified) grass is /bínì/ (Ar.'bonu' or 'bolu' in Dar-Fur) which resembles Pennisetum tiphoideum.[13] It is just as high but the stalk is different; the ears are smaller and the grains are scarce. They look like millet grains. This plant is highly prized but rare. A dense patch of it may be found in the Wadi /oru-ba/ (Kobé). The women harvest the grain to make porridge, the boys to feed the horses when there is a shortage of millet. The stalks are used in making roofs for houses.

/tómsò/ (Oryza breviligulata; Ar. 'am-belele') is wild rice.[14] It is generally eaten in flour-balls seasoned with salt and natron and flavoured with fresh or sour milk. It is commonly found in Bideyat country, rarely among the Zaghawa. In Kobé country it is harvested for the horses.

Cereals gathered during shortages

/nógò/ Ar. 'askanit' (Cenchrus biflorus Roxb.; Gillet, No. 361) offers grains which are edible, but which are in fact only harvested in periods of shortage. These grains bear thorns and are difficult

to gather; the porridge made from them is not particularly favoured.
Nevertheless Bideyat women, with less millet at their disposal, harvest more /nógò/ than Zaghawa women. On the other hand, in the
Sudan, where /nógò/ is not gathered, it has spread considerably at the
expense of other more useful plants. This cereal, however, has the
advantage of being harvested quite late in the year, right up to the
dry and hot season (Z. /āìgì/, Ar. 'sef'), for in bad years, this is
a period when people are generally short of food and very glad to
find /nógò/. The stalks are broken off with a staff, either a piece of
wood or the above-mentioned rake. The ears are beaten with a flail.
From the flour, they make porridge which they eat with milk.

    The grain of /tárà/, Ar. 'drese' (Tribulus terrestris L.; Gillet,
No. 75) too is eaten only in times of shortage The thorny husk must
firstly be removed by crushing them on the grindstone or in the mortar, before grinding the grain to flour. /tárà/ flour, mixed with
/kìè/ flour (see page 20) is used to make a porridge which is eaten
with water or milk. The mixture of the two flours is called /kóndù/.

    These grains are most frequently given to horses. Once burnt,
they are also used to make a kind of tar: /éìgò/ which is used to
treat mangy animals, camels and goats, and even men.[15] As for the
young /tárà/ leaves, they are picked immediately after the rains and
eaten just like /ŋáî/ leaves (see below page 23).

    In conclusion we would like to point out that the majority of wild
cereals we have mentioned, such as /bóù/, the different kinds of
'kreb' or /búbù/, not only replace millet in the preparation of porridge, the basic food, but also in that of beer. This is equally so for
/mádî/ grain, as we shall see later (page 22).

    Among the other possibilities of finding food in periods of shortage we have observed the custom, attributed by the Kobé solely to
the inhabitants of the Dirong and the Guruf, of ripping open ant-hills
to appropriate the grains hoarded by the ants, thus gathering a ready-made harvest.[16] It can attain the size of a camel load of /bóù/ or
/sábà/.

Jurisprudence and food-gathering territories

Do rules of ownership exist for the areas where food is gathered? In
theory one can gather anywhere at all except in the fields. In practice
freedom of action is more restricted. Generally speaking, just as
the herdsman leads his herds back to the same pastures each year,
so a woman returns each year to harvest the wild cereals in the same
spot. She thus acquires, so to speak, rights of use. Should she find
someone else in her stead, she may drive her out. Nevertheless she

has no rights at all over what has already been harvested. So she usually thinks it wiser to go further away, unless the affair is settled by a fight in which the women do not hesitate to use the short knife or reaping hook with which they cut the stalks.

It may happen moreover that an entire village claims a particular wadi for its women, arguing from a right of first arrival, on this occasion at the level of the community. Custom recognises the soundness of such a claim. Yet if one refers to the former position, where clan and village were one and the same, the clan's rights appear pre-eminent.

Another even more imperative rule concerns the herdsmen. They are strictly forbidden to take their herds onto lands known to be rich in wild grasses before these have been harvested. Breaking this rule renders them liable to a fine.

A short ritual precedes the harvest: before going off to gather the wild cereals the women make an offering inside the village. They knead flour with water and make small balls of raw dough which they distribute among the young children. On receiving this food, each child pronounces the following phrase: /anna egeben/ — "May God (Allah) accept [your act]".

Among the Teda, "the harvest of wild grain is also accompanied by certain rites. The valleys where these grasses grow are apportioned in the rainy season among the clans of Tibesti. [...]. Ownership is not exclusive and anyone may come; but what they jealously prevent from happening is that a stranger should be the first to begin the harvest, before the rites may be completed. A man is given the task of keeping the watch on the valley. As soon as the grain is ripe, he warns the people in the clan; these then rush to the spot. The guard of the valley makes a 'sadaga' and throws some 'ediseru' (Ar. 'šii', Artemisia judaica) onto the sandy bed of the enneri. [...]. When the clan to which the enneri belongs has performed its sacred duty and has begun the harvest, strangers are permitted to join in. For this clan, then, there exists only a right of precedence of a religious character".[17]

WILD FRUITS

The classification we intended to use was meant to reflect the user's point of view, to show which wild fruits were most eaten and favoured and which were least so, and to make a sort of list of substitutes offered by wild fruits and plants for sugar, flour, kernels and the young leaves used as ingredients for sauces, etc.

We were most surprised to discover that our order (or rather the native classification thus brought to light) corresponded roughly to a botanical classification of plants; we have accordingly used this latter scheme whenever possible.

Supplementary sugar sources: the Grewia

/kòrfù/ (Ar. 'tomur el abid', "the slave's date") is the fruit of a shrub, the /kòrfúʀà/ (Grewia villosa Willd.; Gillet, No. 105). The fruit ripens in September-October shortly before the millet harvest. Women, children and herdsmen pick it off the tree.

Both children and herdsmen suck it on the spot, first carefully removing the skin by squeezing the fruit between their fingers and spitting out the four hard seeds. One only eats the sweet flesh. Occasionally one may eat /kòrfúʀà/ flowers.

Women gather the /kòrfù/ and store it. The dry fruit looks like a little brown seed and has a sweet taste. One extracts sugar from it in two ways: by soaking it in water for a day or by boiling it, which speeds up the process. Women mix the syrup thus obtained with flour to make a sort of porridge specially eaten during Ramadan.

Fruits that have received this treatment are not thrown away but are used in due course to make tar, as with /tárà/ grain (see page 17).

/kòrfúʀà/ wood is used to smoke homes in order to rid them of bad smells and flies during the rainy season.

/ňáʀi/ (Ar. 'giddem') is the fruit of the /ňáìʀà/ (Grewia populifolia Vahl.= G. Tenax Forssk.; Gillet, No. 194),[18] a shrub from the same family as the previous one. It is sweeter than /kòrfù/ and thus more appreciated; but in the country as a whole more /kòrfù/ is found than /ňáʀì/. Both ripen at the same time after the rains.

Just as with /kòrfù/, children eat flowers and fruits, this time without throwing away the seeds, which they chew. Women gather and store the fruits from which they extract sugar by steeping them in cold water or boiling them. The syrup thus obtained can replace sugar in tea or sweeten porridge made from millet, 'kreb' or 'absabe'.

/gúgúr/ (Ar. 'kabayŋa' [a Maba loanword]) is the fruit of the /gúgúrdà/ (Grewia flavescens Juss.; Gillet, No. 103). The fruit is eaten raw or else its sugar is extracted to sweeten porridge.

/sóŋò/ (Ar. 'baxšem' or 'baxšamay') is the fruit of the /sóŋóʀà/, an unidentified tree; our informants regard it as belonging to the same family as the /gúgúrdà/. Its fruits are picked and eaten in the same way as the /ňáɳ̀/; the seeds are chewed. The wood of this tree is used for making spear shafts.

Three out of the four fruits which we have just discussed belong to the Grewia species — perhaps the fourth does too; eaten fresh by children and herdsmen, they are used dried by women who preserve them in their granaries. According to some, /sóŋò/ and /gúgúr/ are the most popular fruits. From each of them is extracted a sweet syrup used to sweeten the various porridges eaten in the month of Ramadan. During this period the meal taken after sunset, which breaks the fast, is prepared with special care and the serving of sweet foods is evidence of its festive character.

Supplementary sources of flour and sugar

i. The jujube tree — /kîè/ (Ar. 'korno') is the fruit of the /kêɪRà/ (Ziziphus mauritiana Lam.; Gillet, No. 205).[19] It is a fruit that grows abundantly and is greatly enjoyed. It consists of three parts: the pulp /kóndù/, the stone /míɲà/ and the kernel /míɲà bùr/. The fruits ripen around March-April. They can be eaten freshly picked: the children and shepherds eat them without stint, the adults likewise when seated in a group chatting. They eat the pulp and collect the stones which they proceed to crush between two large pebbles to extract the kernel.

Women use them in cooking after preparing them. When the pulp has been dried and pounded in a mortar, they obtain a flour which will be eaten mixed with millet flour, in the form of porridge or boiled in water or milk, or as cakes. To make the cake, a hole is dug in the ground; it is filled with /kîè/ flour, then covered over with branches, leaves and cow-dung, which are set alight. When these have finished burning, a very hard cake, which is eaten dry or together with milk, is taken out of this improvised oven.[20]

Using the kernels and the sugar extracted from the fruit, the women make an excellent nougat, /míɲà ásán/ (from the Arabic 'asal', "honey"), which they eventually sell in the markets. It is made in the following way: jujubes are boiled in an earthenware vessel over a fairly long period, the sugared water is collected and the stones are placed in a hole where they are sprinkled with water at frequent intervals for two days; this latter operation is to prevent the kernel from being crushed into a pulp, when it is extracted by breaking the jujube stone between two big pebbles. The kernels are then thrown into the boiling sweet syrup, which is taken off the fire when it has reached a burnt caramel colour.

Lastly the tree's wood is used for rafters.

/kábárà/ (Ar. 'nabak') is the fruit of another jujube tree, /kábáɪ̀Rà/ (Ziziphus spina-christi (L.) Willd.; Gillet, No. 206). This fruit, smaller than /kîè/, offers the same edible parts and is similarly used. It seems to be preferred for its extra sweetness. The smoke of /kábáɪ̀Rà/ wood is used to scent houses and clothes.

ii. Balanites — /gîè/ is the fruit of the /géyRà/ (Ar. 'hejlij'; Balanites aegyptiaca (L.) Del.; Gillet, No. 209). It looks like a small plum and it consists likewise of three parts: the pulp, the stone /gógúrù/, and the kernel /góróù/ which is bitter. These fruits may be eaten green or ripe.

The green fruits /gêì-mè/ are very bitter; they are never eaten raw but cooked in water. After cooking, they are not eaten hot for fear of dysentery, but only after being spread to cool outside. It is used as a food in times of shortage.

The ripe fruits are eaten raw. As with the Grewia fruits, they may also be boiled in water for a whole day to extract the sugar; the syrup produced is used to sweeten millet porridge.

From the kernels the women make nougat. However before this, the bitter taste must be removed. This is done by boiling them in water, which is changed three or four times, then placing them in an earthenware jar filled with cold water, where they are left for three to four days. They are then dried in the sun; the kernels are thrown into the boiling syrup.

/gîè/ berries, kernels and nougat are kept in granaries and used whenever needed. In the past oil was commonly extracted from /gîè/, but nowadays this technique appears to have been given up and to be almost completely forgotten.

The unripe fruit may be used to bleach /tógúí/ cloth. The pulp is crushed between two stones and used to rub the material, which is then beaten with sticks.

Lastly the blacksmiths manufacture mortars, pestles and stools from /géyRà/ wood; and the herdsmen hew themselves curved throwing sticks.

/génè/ (Ar. 'himed'), the fruit of the /généRà/ (Sclerocarya birrœa (A. Rich.) Hochst.; a tree identified by H. Gillet) is also eaten. It looks like a yellow plum and has a thick skin. Its flesh which is very juicy but tasteless is edible and the stone is broken to remove the kernel. This tree, scarce in Zaghawa country, grows mainly in Tama and Maba countries, more to the south.

Cereal substitutes

i. The Capparidaceae — /mádì/ (Ar. 'moxet') is the fruit of the /mádíRà/ (Boscia senegalensis (Pers.) Lam. ex Poiret; Gillet, No. 6). It is a round, rather hard berry, the size of a small cherry with two seeds joined together inside. /mádì/ designates the green unripe fruit; the ripe fruit is yellow; its name is /ayar/.

When the fruits are ripe, before the rainy season, the herdsmen suck them to swallow the juice and they spit out the seeds. It is however these same seeds that the women use as a food in the following way. When the fruits are picked green during the cold season, the flesh is removed in order to take out the seeds; when picked ripe, they are dried in the sun, then crushed to remove the skin. After this the preparation is very much the same in both cases, perhaps slightly longer in the case of the green fruits, which are more bitter. The seeds are quite thoroughly washed, then put to soak in water which is changed each day after squeezing out the juice. In the case of green fruits it takes a good week to remove the bitter taste. As a rule women perform this task beside the wells for more convenience. The seeds are then put to boil in water which is changed three or four times. A little natron is added, as it has the property of softening them. After this lengthy preparation (which requires a fortnight for green fruits), the women make a porridge which is eaten with milk or clarified butter.

The seeds put aside in the granaries are prepared as a substitute when one is short of millet.

As we have already pointed out, women can also make 'merise' from /mádì/ grain. But it is a beer of very inferior quality and it is an insult to tell someone: "You drink 'merise' made from /mádì/".

Donkeys like to eat the bark of the /mádíRà/; and the wood of this tree is valuable in the rainy season because it is easily ignited, even in the rain (just as with the branches of the /námárdà/, discussed below).

We would like to point out two fruits in the same family as /mádì/ eaten only by children and herdsmen: /námár/ and /núr/.

/námár/ (Ar. 'tumtum' or 'tundub') is the fruit of the /námárdà/ (Capparis decidua (Forssk.) Edgew.; Gillet, No. 9; or C. sodada ?). It is eaten ripe by herdsmen and children, but is not generally harvested by women and is therefore not brought home.

/núr/ (Ar. 'kurmut') is the fruit of the /núrdà/ (Maerua crassifolia Forssk.; Gillet, No. 19).[21] It is eaten ripe by children and herdsmen.

The leaves of the /núrdà/, of great nutritive value, are given to

horses and the thin branches are used for making rakes to gather
wild grasses.

A week after the rains, the leaves of a plant with a taproot are
gathered, while these are still tender; the plant is called /ŋâî/ (Ar.
'timilexe' or 'timileiki' identified by H. Gillet: Gynandropsis gynandra (L.) Briq.) and belongs to the same family as the previous plants.
These leaves are bitter; they are boiled for a long time with frequent
changes of water. Salted and seasoned with clarified butter they make
a dish that everyone eats as a substitute during the period when
cereals are short.

ii. The Cordia — /túrù/ (Ar. 'andarab') is the fruit of the /túrdằ/
(Cordia Rothii Roem. et Schult.; Gillet, No. 255; or C. gharaf ?). The
fruits come in short clusters; they are round with a stone inside.
When ripe, they are a reddish-orange colour and turn reddish-black
on drying. They are eaten raw and are completely edible. When
found in profusion the women collect and dry them, then store them
away in their granaries. In times of shortage they are used for the
preparation of porridge.

/túrdằ/ branches are utilised to make the circles supporting
house rafters.

iii. The Commiphora — The stones /tógóʀîằ/ found inside the drupes
of the Commiphora africana (Ar. 'gafal'; Z. /tógórò/; Gillet, No.
210) are also eaten. They produce a flour which is mixed with that
of millet to make porridge. The latter is then consumed together
with milk. Apart from this, the seeds are crunched raw by the
children. They may be used as well to make tar.

Tubers

/nógù/ (Ar. 'siget'; Cyperus rotundus L. subsp. tuberosus; Gillet,
No. 341) is a plant which appears at the beginning of the rainy season.
The subterranean tubers found right at the very end of the roots are
eaten. They may be eaten just as they are; they are black, just like
grains of pepper; their slightly sweet taste is somewhat acrid and
floury. Women offer them roasted or crushed and cooked (in which
case they are presented in the form of small balls) to buyers in the
markets.[22]

There are two kinds of /nógù/: /nógù bei/ "goat nógù" and /nógù keni/
"ewe nógù". The first kind grows on the banks of the wadis, in the
sand, and its tubers are of a whitish colour. It grows scantily and
is harvested by young children.

/nógù keni/ grows around pools, on clay soils. Young women go off for a month, sometimes two, to gather them. This is done by beating the ground with a flail, thus softening it, and the young women simply have to gather the tubers on the ground and sift them.

Other fruit

i. Salvadoraceae — /ùí/ (Ar. 'šao') is the fruit of the /ùíRà/ (Salvadora persica L.; Gillet, No. 201). These fruits have the same appearance as those of the /túrdà/ but are smaller. They are consumed raw.

Cut into short sticks, /ùíRà/ branches are used for cleaning teeth.

ii. Cucurbitaceae — Water-melons (Z. /órù/, Ar. 'battix'; Colocynthis citrullus (L.) O. Ktze; Gillet, No. 90) grow wild almost everywhere; in the Guruf and in the regions of Anka and Dor they are grown in millet fields too. They are eaten either raw or else boiled or roasted in ashes. Their seeds are enjoyed when roasted and can also be used to make tar.

The colocynths (Colocynthis vulgaris Schrad.; Gillet, No. 91) are eaten in the same way.

/túdù/ is the fruit of a large liana, /túdùRà/ (Coccinia grandis L.; Gillet, No. 89) which creeps along the ground or frequently climbs on trees. This plant loses its leaves in the dry season and grows in the rainy season. Its fruit, elongated like a cucumber but shorter, is fairly big and contains a large number of seeds. When ripe it is soft, its colour red and its stem green. It is very sweet and juicy. Herdsmen and travellers eat it.

After being beaten with sticks to remove the bark and then softened, /túdùRà/ stalks are used as string.

iii. Malvaceae — Among the Hibiscus, one small shrub offers quite remarkable possibilities. It is the Hibiscus sabdariffa L.; Gillet, No. 118 (Z. and Ar. ?/áŋárá/ also /kerkere/ or /kerkedi/ in the Sudan). Its leaves (Z. and Ar. /kárkáñ/) are eaten after boiling. When seasoned with oil, this dish with a slightly acid taste like that of sorrel is eaten with millet porridge. Also consumed are the seeds /kómùn/ contained in the fruit which is a capsule. They are eaten roasted or else turned into flour from which porridge is made. They may also be used to make tar. Lastly a beverage, which often replaces tea, is commonly made from the red sepals /áŋárá/. It is drunk very sweet, retains its acid taste and is very refreshing. The dry sepals are sold on the market. Traders even export them in large quantities to pharmaceutical laboratories in Germany.

This plant which grows wild is occasionally sown in millet fields, as we frequently saw in the Sudan.

/áŋárá/ belongs to the same family as okra (Hibiscus esculentus; Z. /nyaʀi/), a cultivated plant.

A few medicinal plants

/médèr/ (Ar. 'ardeb') the fruit of the /médèrdà/ (Tamarindus indica L.; Gillet, No. 144) has the form of a pod. The seeds may be eaten but their taste is very acid. It is used essentially for medicinal purposes, and pods are gathered and kept in case of illness. A sick person is given water to drink in which pods of /médèr/ have been soaking.

/médèrdà/ leaves are also boiled to wash children who have had chicken-pox.

The pods /bírgè/ of the /bírgéʀà/ (Ar. 'garad'; Acacia scorpioides (L.) A. Chev.; Gillet, No. 149) are used to stop haemorrhages. The whole pod is used, either ground to powder or chewed, after which the mixture is spat on the wound.

The pods of /bírgéʀà/ are also used for tanning skins.

The sap of the Calotropis procera Ait. (Gillet, No. 217); Z. /kõrfú/; Ar. 'ušar', a poison, may be used as a medicine to treat spots; the spot is scratched until it bleeds and a little of this "milk" is applied to it. It is also an abortive remedy often used by women expecting illegitimate children: they drink half a glass of /kõrfú/ "milk".

An infusion with which eye-sores are treated is made from the seeds of /ósù sũlí/; Ar. 'fileya' (?) (Ocimum hadiense Forssk.; Gillet, No. 317).

OTHER PRODUCE FROM GATHERING

Apart from seeds and wild fruits, the Zaghawa include other produce from gathering into their diet; some of these may be eaten by young children only, but other foods that could be collected are regarded as prohibited.

People collect the eggs of guinea fowl or ostriches scattered about the bush. Everyone eats guinea fowl eggs hard-boiled or cooked in the ashes. As for ostrich eggs, though it appears that everyone may eat them, we have never seen them eaten except by old people who usually sucked them. But only children may eat the eggs of other birds.

Mushrooms, snails, worms and frogs are not eaten, but locusts are much appreciated. Of these the red ones are preferred to the black ones; young locust hoppers /égímè-bur/ are especially sought after. They are easier to catch, which is done by digging a hole in front of them. The locusts fall into it; and one only needs collect them or else light a fire in the hole to kill them. They are boiled in salty water or fried, and eaten. Large locusts, too, are caught when numb with cold. In this case they are burnt by means of a long stick with a tuft of burning grass at the end.

Only uncircumcised children eat young birds or hares; however blacksmiths may also eat hares.

Everyone, of course, collects wild honey when found.

This attempt at an inventory of the products of gathering gives only a limited insight into their many uses. We cannot bring it to a close without mentioning the use of grasses for the manufacture of basketwork, as with the Eragrostis cilianensis (All.) Link ex Vign. Lut. (Gillet, No. 380; Z. /mine/) and the use, similarly, of barks, in particular that of the Acacia seyal, in the manufacture of both basketwork and carrying nets. Also calabashes as containers for milk and butter. Gum from the Acacia Senegal (L.) Willd. (Gillet, No. 150; Z./túè/; Ar. 'kitir') is utilised in various ways: sweets, medicine, stopping leaks in receptacles. A traditional export of the Sudan, it has only been tapped as a cashcrop in the Chad for the last twelve years or so.[23]

Browsing opportunities are provided by that same Acacia seyal, whose leaves and pods are knocked down when there is a shortage of grass (see page 35). Various leaves, picked when very young, constitute one of the ingredients of sauce, such as those of a plant belonging to the Grewia family, which grows in sand: Z. /múlúkíé/ (from the Arabic 'muluxiya') Corchorus olitorius L. (Gillet, No. 100), those of the /ŋáĭ/ (mentioned above) or else the leaves of /káwán/ (Ar. 'kawal') Cassia tora L. (Gillet, No. 142), with a sorrel flavour, which women prepare by pounding them to extract the juice and leaving them to ferment in a covered pot for a fortnight before drying them in the sun. They use them when necessary or go and sell them in the market. The seeds of the latter are also of use. A kind of tea, which is drunk as a remedy for head-aches, stomachaches, fatigue or any unidentified illness, is made from them.

Lastly there is the wood provided by the Commiphora africana (Z. /tógórò/; Ar. 'gafal'), the Cordia Rothii (Z. /túrdà/; Ar. 'andrab'), the Acacia seyal (Z. /musumara/, Ar. 'talha') for the manufacture of kitchen and domestic utensils: dishes, stirrers, stools, flails and the bodies of musical instruments, drums and harps.

As an economic activity, gathering is an exclusively female occupation among the Zaghawa — leaving aside the young shepherds, who satisfy their appetite or greed depending where their wanderings take them — and every woman, be she princess, 'miskīn' (poor) or blacksmith's wife, is a gatherer at some time or other. We would however like to remark that we noticed men gathering wild grasses with the intention of supplementing their horses' feed.

The position among the Teda is identical; there, "gathering is a normal, almost continuous activity, each season contributing its own produce" and the harvest of wild cereals "is the occasion for real female expeditions".[24] Apparently, there too, whether they be Teda or Daza, or else 'duudi' for the Teda, 'azza' for the Daza (i.e. blacksmiths' wives), all the women take part in this activity.[25] As for the participation of men, according to J. Chapelle, it is the man who sometimes travels over 200 km to gather a supply of wild colocynth seeds,[26] though Charles Le Coeur considers this task to have also devolved on women, who alone go and pick fresh fruit, just as with all seeds.[27]

The picture presented on the one hand by the whole of the group (Teda, Daza, /bèRí/) and on the other by a large number of peoples of the Eastern Chad is one of great homogeneity. Whether fixed, semi-nomadic or nomadic societies are concerned, whether the principal activity is husbandry or the cultivation of millet, women everywhere devote a large part of their time to the gathering of seeds and wild fruits. From east to west the same is true of the Kanembu, the Kuka, the Murro, the Dagal and the Kibet, the Masalit and, towards the north, the Maba, the Mararit and the Tama, without mentioning the numerous nomadic Arab tribes.[28]

This situation, comparable to that of the Nigerian Sahelian zone where the Tuareg also go in for the harvesting of grasses and wild fruits, differs from it in one essential respect which deserves to be noted here. In actual fact, among the Tuareg, men intervene in the gathering "for tasks which require a strenuous effort or when gathering is carried out on a large scale". An even more remarkable feature is that the gathering of wild plants in this society is the task of the serf caste, the Iklan, who once provided for their Tuareg masters and now sell them the product of their harvest.[29] The difference is more apparent than real, if one considers the fact that it is not unusual for male slaves to perform female work. Only two conclusions are thus possible: either the Tuareg practised gathering as a female activity and discharged it through their slaves or else they were unaware of it and heard of it through their black slaves.

Such then are the quite diverse and relatively abundant resources which a poor country can offer to those like the Zaghawa who possess a sufficient knowledge of the natural potentialities to be able to exploit them.[30] Plants and trees, seeds, fruits and leaves are known and classed, not by a small number of specialists but by the whole population. These "wild" foods are the means by which each person makes up food shortages and survives. In most cases the fruits can be eaten in the fields but much of the time, just as with cereals, they are stored in granaries and used according to need and according to the season. Part of the products of gathering are put on sale and in the markets one may see women selling them, where available, to other women. This restricted trade, or rather barter, takes place exclusively between women, whereas the trading of millet is the business of men (except where, as with gathered produce, very small quantities are involved).

Were the agricultural potential to improve or the commercial exchanges to intensify and the demands for food to be satisfied on the whole, as is the case in the Sudan, then the activity of gathering, which absorbs a considerable amount of time both in the collection and in the preparation of the produce harvested, would disappear; but at the same time the knowledge which the users had of the wild plants would decline.[31] We have observed moreover that in the areas where gathering is no longer practised, the least useful plants occupy the land at the expense of the better ones. We feel that one must not lose sight of the risks inherent in agriculture in this region nor underestimate the danger represented by a complete abandonment of the techniques and traditional methods of gathering.

If gathering may require a longer time for the collection and preparation of the foods, one must not forget that, against this, the users, whether men or women, are spared the tasks of preparing the ground and sowing, of weeding and watching the fields. Cultivated cereals, whose seed has been entrusted to the soil at some risk, will equally need some form of preparation, after their harvesting and storage in granaries, before they can be turned into food.

Are there any other reasons to explain why the practice of gathering, still current among the Zaghawa of Wadday, has been currently abandoned by those of the Dar-Fur? Besides those reasons previously noted and accepted: better agricultural potential and greater commercial development, one must stress another, of a sociological nature, that was clearly explained to us by a Zaghawa informant from the Dar-Fur: "The people of the Chad gather seeds and wild fruits because they don't have enough millet and don't want to go and buy some in the south. They keep their money to acquire animals in order to get married". In fact the bridewealth paid to obtain a girl in marriage is far higher among the Zaghawa of the Chad, where it can attain, they say, a hundred head of cattle; among the Zaghawa of the Dar-Fur a girl may be obtained for twenty cattle. This is why the Zaghawa Kobé, among whom the investigations about gathering were chiefly conducted, are so reluctant to part with their currency but acquire millet by barter rather than by purchase, and complete their requirements of food by obtaining the necessary supplement through their own gathering.

Original French article, 1969: "La pratique actuelle de la cueillette chez les Zaghawa du Tchad", Journal d'Agriculture tropicale et de Botanique appliquée, 16(2-5): 55-83.

Chapter 3

THE PASTORAL SYSTEM AND THE NEED FOR TRANSHUMANCE

The tendency to give up wild grain gathering for cultivation wherever the conditions make it possible is in a sense equivalent to discharging the women of their burden and sharing it with the men, who have certain tasks to perform in the fields. One might furthermore remark that this tendency is probably encouraged by the fact that a diet based on millet is more valued than one composed of wild produce. Besides, and we have had no opportunity to discuss this with our informants, we somehow feel that for the Zaghawi the cultivation of millet represents a higher standard of culture. A parallel might be drawn here between the relation of cultivation to gathering and that of animal domestication to hunting. In either case the activity implying a greater interference of man with natural processes is the more valued and its produce more highly considered. One cannot avoid being struck by the fact that gathering is an exclusively female activity, in the same way that hunting is only practised as a profession by the despised blacksmiths.

It would be no distortion of reality to state that, among the Zaghawa, the man is a herdsman or part-time farmer, who only goes occasionally hunting (and nowadays preferably with a rifle), whereas the woman is a food gatherer and the blacksmith (who until recently neither raised cattle nor practised cultivation) is primarily a hunter, using nets and spears. Even in the Sudan, where millet fields are extremely popular, the Zaghawi's first thoughts nevertheless go to his animals.

Like most herdsmen in the area, the Zaghawa are faced with the problem of raising large flocks and herds, for they would like them to keep on increasing, in a country with scarce water-supplies and pastures. To cope with this situation, they employ their restricted resources in as rational a fashion as possible: after the rains they dispatch their herds far into the more arid north to consume the new grass that will soon wither under the sun, drinking at the temporary pools; by this means they keep in reserve the pastures close to the permanent wells around which the animals will gradually congregate during the dry season. A two-way movement is thus involved, consisting of an expansion outwards — which begins with the rains and extends into the dry season till the end of December — and of an inward contraction during the following six months.

THE SEASONS

The Zaghawa recognise three distinct seasons: /gyé/, /dábó/ and /áìgì/, to which must be added the period of harvest, /tàrbà/, and two brief intermediate seasons surrounding /gyé/.

The annual cycle begins with the rainy season, /gyé/, preceded by a short period, /ìrsãsi/ (Ar. 'uršãš') or /bógóù/, in the course of which appear the first clouds heralding the arrival of the rain.

The rainy season proper, /gyé/ or /íɒí gyé/: "wet earth" (Ar. 'xarīf'), begins about June, sometimes May, and comes to an end in late August. The temperature then is relatively cool (mean maximum 35^0 C, mean minimum 18^0 C). At first infrequent in June-July, the showers may become daily in August, although the total precipitation may also be attained in a single shower. The annual total never exceeds 600 mm and varies generally around 350 mm: 374 mm at /hiʀi-ba/ (Chad) in 1954. In the same village in 1957 we observed the following levels of rainfall:

May	10.4 mm in two showers
June	21.2 mm in four showers
July	98.1 mm in six showers
August	455.2 mm in eleven showers, including one of 94 mm on the 25th August
Total	584.9 mm, which was a very good year.

But the region of Hiri-ba, situated in the centre of the western Zaghawa plateau at a height of nearly 1,000 m (935 m at Hiri-ba), is better watered than for instance the regions of Anka, Musbat or Forawiya (in the Sudan), that border onto the desert and where totals vary around 150 mm.[32]

Reproducing the figures collected by C. McRamsay, Pierre Quezel gives these figures for Kutum, a centre situated in the south-east, on the border of Zaghawa country (Dar-Fur):

1929-1949	351 mm (annual mean)
1964	266 mm
1966	294 mm

For Um Buru, in /tuer/ country, he provides this series of measurements:

1958	189.5 mm	1963	205.5 mm
1959	269.5 mm	1964	263.5 mm
1961	516.5 mm	1965	296.5 mm
1962	268 mm	1966	265.5 mm

which works out at an average of 284.5 mm for the eight years concerned. If one examines the map drawn up by P. Quezel from the conflicting observations of C. McRamsay and J. G. Lebon, the Sudanese part of Zaghawa country is clearly situated between the 300 and 100 mm isohyets.[33]

A second short, intermediate season, /mùgùlí/, brings the rainy season to a close.

Then follows a dry and hot season: /tằrbằ/ (Ar. 'darat') which corresponds with the period of harvest and extends through October.

Next, the temperature grows cooler and this means that /dằbó/ (Ar. 'šitā'), the dry and cold season which lasts from November to January has arrived. Nightfall is marked by the onset of a sharp chill which persists well after sunrise and in December it is common to see the thermometer go down to 4 or 5 degrees centigrade. A fierce wind blows from the northeast and a completely overcast grey sky shuts off the sun for days on end.

With February comes the dry and hot season: /ằigì/ (Ar. 'sēf'), which lasts till June. It is a critical period, when the first rains are impatiently awaited; sometimes a few showers fall in May, sufficient to revive the vegetation, so helping to bridge the gap.

Thus the shepherds have to endure at least nine months of dry season before reaching a short rainy season of such irregularity that several years of total drought may follow in succession. In some years the total rainfall is inadequate; in others, while the total may be fairly high, the showers are not sufficiently spaced out to allow the pastures to grow again.

But abundant falls of rain spaced out at intervals may cause the pastures to extend their limits much further north into the desert. When the rain is scanty or the intervals too short, the usable area correspondingly diminishes. In that case the herds make for the south into foreign parts thus raising the awkward problem of coexistence with the resident cultivators there. For just like the nomad herdsmen the latter have animals to feed and water, at the very moment when they see their crops suffering from the same drought.

VEGETATION

With an average altitude varying between 800 and 1,000 m, the first aspect of Zaghawa country is that of an immense steppe of grasses and thorns. But throughout its extent, exceeding 400 km from east to west and almost as much from north to south, there is a very great variety of landscapes and soils to be found.

The traveller discovers vast sandy stretches in the region of Anka (Sudan) or in the Guruf (Chad); elsewhere, crystalline plateaux swept by the east wind, from which rise rocky peaks like /trĭŋè/ (given as "Taringel Rock" on maps, north of Um Haraz), a sharply outlined plateau ledge, like the Kabka, resembling a mountain range, small hills split by erosion into a number of precariously balanced blocks like the heights of Mir and Kobé in the Chad.

During the rainy season the main wadis flow for a few hours, or two or three days at most. Water pours into the clay hollows turning them into immense pools. Everywhere the ground is overrun with green grasses that will wither quite quickly as soon as the rains cease. Then the most common colours will be the vivid yellow of the pastures and the lighter yellow of the fine sand belonging to the dried out wadis. Against a blue sky swept clean by the east wind stand the dark green silhouettes of large acacias, which sink their roots deep in the beds of the watercourses or in their banks in search of the underground deposits of water. The sparser foliage of other thorny trees growing on the plateau traces lighter shapes in which the pale green of the tiny leaves is set off against the white, long sharp thorns.

The fine sandy soil of 'gōz'[34] (Z. /šige/) favours the cultivation of millet and the proliferation of wild grasses. The clay soils (Z. /márdĭ/, Ar. 'naga') of the flat and uniform hollows, which were covered over by water during the rainy season, now in the dry season take on the appearance of a cracked and deeply fissured brick pavement. The Zaghawa call them /ādĭ/ (Ar. 'berbere'?). Often they are strewn with pebbles and then called /kárákárá/.

The vegetable cover consists for the most part of grasses of the following types: Aristida, Eragrostis, Brachiaria deflexa (Z. /ègé/, /áìrì/, /jígĭʀì/; designated together by Z. /gú/, Ar. 'kreb'), Panicum (Z. /sábà/, Ar. 'kreb')[35] and Dactyloctenium (Z. /bóù/, Ar. 'absabe'). In their Catalogue raisonné et commenté des plantes de l'Ennedi, G. Carvalho and H. Gillet identify fifty types of grasses[36] and in his above-mentioned study of the vegetation of northwest Dar-Fur, P. Quezel lists over a hundred. Whether perennial or annual, these grasses grow with the rains and remain green for a few days only. Some of them disappear under the action of the wind and others survive as straw. They offer the animals green fodder during a short period and dry fodder throughout the rest of the year.[37] Besides, their seeds gathered by the women provide a supplement of cereals for human consumption.[38]

The trees found most widely are acacias of which the finest without any doubt is Acacia Faidherbia albida (Z. /teli/; Ar. 'haraz') whose favourite spot is the banks or even the middle of wadis. Thus in the middle of the wadi Koba an enormous /teli/ overshadows the permanent wells found there; it is the same at Oru-ba, Bir tawil and in many other places. "When we find /teli/ in a place, we dig wells and we are sure to find water" say the Zaghawa.[39] Botanists agree with the local people in making this tree an indicator of underground water: they have observed that it "is always linked to an underground deposit of water by a very long tap-root".[40]

Among the other kinds of acacias, we should particularly like to mention Acacia scorpioides (Z. /bírgérà/; Ar. 'garad'), Acacia seyal (Z. /musumara/; Ar. 'talha'), Acacia senegal (Z. /túè/; Ar. 'kitir'), all of which prefer the banks of wadis or the large pools, and also Acacia raddiana (Z. /kedira/; Ar. 'seyal'?) which grows on the hard and sandy soils of plateaux, just like the Acacia senegal.

Ziziphus mauritiana (Z. /kêìrà/; Ar. 'korno') and Ziziphus spina-christi (Z. /kábáìrà/; Ar. 'nabak') together with Balanites aegyptiaca generally prefer the banks of wadis. Commiphora africana (Z. /tógórò/; Ar. 'gafal') manages to cling to rocky places. It grows beside a few Acacia mellifera, Grewia flavescens (Z. /gúgúrdà/; Ar. 'kabayŋa'), Grewia villosa (Z. /kòrfúRà/; Ar. 'tomur el abid') and Boscia senegalensis (Z. /mádírà/; Ar. 'moxet').

Almost all these trees provide browsing pasture for camels. While humans eat their fruits,[41] camels browse on the pods and the leafy branches, in spite of the thorns for much of the year. Furthermore some of these offer good pasture for goats, like Acacia seyal and Maerua crassifolia (Z. /núr/; Ar. 'kurmut') whose fruits and upper leaves are knocked down with a specially made hooked stick /kî/, by shepherds, for even standing on their hind legs, goats cannot reach them. In April-May only the trees still bear some greenery. Thus besides the vegetable cover, there exists a second form of pasture for camels and goats, usually described as "browsing" which allows them to survive during the dry season.[42] Moreover the camels can await the arrival of rains due to the fact that numerous species, like Balanites aegyptiaca for instance, become green again in March-April.[43]

The various possible uses of these many kinds of pastures are well known to the local population which values their distinctive qualities such as coarse or tender, more or less salty, constipating.

Lastly we have to consider 'jizu', a particular kind of pasture covering sandy sparsely watered areas (under 200 mm), especially to the north of Dar-Fur. Michel Baumer, an ecologist, variously defines 'jizu' (from the Arabic 'jaz': to live without water by drinking the sap of plants) as 1. a pastoral practice, 2. the area where it is in use and 3. the type of vegetation found there. The practice is followed by the Kababish and Kawahla of Kordofan, the Midob and Zeyadiya of Dar-Fur (to whom one should add the Zaghawa) and the Gur'an of Chad. Between October and February these peoples take their camels and adult sheep together to the eastern border of the Libyan desert — provided rain has fallen within the past two years in the area where they intend to live as nomads far from any watering point. 'Jizu' territory extends from the northeastern horn of Kordofan, gradually widening towards the vicinity of the Ennedi.

'Jizu' vegetation is of a desert type: short lived, its growth is ensured by nocturnal falls of dew. It is an association of a few plants, which are, in order of decreasing importance: Indigofera bracteolata (Ar. 'derma'), Neurada procumbens (Ar. 'saadan') Indigo arenaria (Ar. 'xušēn'), Triraphis pumilio (Ar. 'saleyan'), Aristida papposa (Ar. 'missa'), Crotalaria thebaica (Ar. 'nataš'), Tribulus longipetalus (Ar. 'guttub') and Fagonia cretica (Ar. 'agul'). 'Jizu' appears after the rains have fallen further south and the nights have become chilly, i.e. October or November.[44] As regards the region we are interested in, P. Quezel defines the 'jizu' association, in its southern occurence at least, as a group characterised by Schmittia pappophoroides and Panicum turgidum, setting its southern boundary 4 km north of Musbat. He stresses the "nature of the substratum, which here appears to retain more easily a sizeable proportion of humidity in its upper layers ..." and "the presence of hardy grasses and cyperaceae, together with the decline in Cenchrus biflorus, allowing a prolonged use".[45]

Zaghawa herdsmen are well acquainted with these conditions and around November, sometimes as early as October, they send their herds of camels and flocks of sheep to the 'jizu' to spend the dry and cold season /dábó/ there. According to our informants, certain grasses are eaten by camels, others by sheep. Hence the general practice of putting these animals together in the groups making for the 'jiźu'. The practice has been pointed out by M. Baumer. It would however be interesting to know what kinds of plants are chosen by camels and sheep respectively and for what reason this choice is made. P. Bourreil has attempted to provide an answer to this question (see page 95).

THE WATER SUPPLIES

Pools[46]

In the rainy season small pools, which vanish almost as quickly as they appear, are everywhere to be found. Although temporary, they play an important role in the pastoral economy, for their wide dispersal means that the animals are never very far from water.

However the largest pools, which fill the clay hollows and retain water for six months or more, serve a quite different purpose. The main ones are the pools of Um Shedar, between Kutum and Mellit in the far eastern corner of Zaghawa country, of Jinnik, north-west of Anka, and of Key-hay (given as "Matadjéné" on French maps) at the foot of the Kabka range in the far west corner of Zaghawa country. Later the water gradually recedes from these vast expanses, dotted with acacias, leaving in the dry season a hard crackled earth, in which the animals' hooves have left deep marks, as hard as baked brick.[47] These places continue to be visited, even when the pools are dry. In such a case water is drawn on the spot from wells which have the advantage of being permanent.

The lake of Umdur, on the border between the Chad and the Sudan, may be regarded as permanent, though according to some informants it has dried up in certain years.[48] With a view to storing water two dams have been constructed by the Sudanese government since independence, one at Ba-sao in Kobé country, the second at Metel-koru in Tuer country south of Um Buru.

Wells

To obtain water the Zaghawa dig holes (Z. /kŭná/, Ar. 'tamada') somewhat at random in the wadi beds. As they are shallow, such rough and ready wells require no timbering. A calabash is used to draw the water, which oozes onto the sand at the bottom. Such draining wells serve during the rains and afterwards. A dozen or so are often to be found in close proximity. As the water level goes down, wells are gradually deepened. This stops when the well has become too deep and is in danger of collapse.

The permanent timbered wells (Z. /bà/, Ar. 'bir') generally attain a depth of between 15 and 20 m. Some may reach 60 m. Water is drawn with a kind of dipper (Z. /kwòí/, Ar. 'dalu'), attached to a rope.

The French, English and (following independence) national administrations have built cemented wells where as a rule water is drawn by hand, just as with the local wells.[49] A stone or brick curb prevents the water from being polluted by the mud and animal excrement around the well. Unfortunately however the dipper (or tin can occasionally found in its place) is always left on the ground before being let down to the bottom. In addition to the well there are often watering-troughs built in the same fashion.

Timbered wells are the work of the Zaghawa themselves. A varying amount of care goes into their construction; it depends above all on their depth and no doubt also on the degree of professional competence of the expert (formerly always a blacksmith) who directs the work. For instance at Anka, although the water is at a depth of 4 or 5 m, the three wells dug in the wadi were merely reinforced at the top on all four sides with timber. All around, the bed of the wadi was thickly dotted with collapsed wells, dry during this period (November), their timber whitened and scored by the movement of the ropes. At Ar-koila, at the foot of jebel Ha-yei, a well reaching a depth of thirty men, as we were told, in other words about 60 m, offered a magnificent specimen of craftsmanship with its sides fully timbered and its narrow lip-shaped opening, to prevent people and animals falling in. In fact it is not uncommon for a cow to fall to the bottom of a well, especially when the latter is a simple hole dug in the sand with sides liable to subside. The animal is pulled out with the aid of ropes.

Every year the wells dug in the wadis (Koba, Anka, etc.) or else a few metres away from their banks (Hiri-ba, etc.) must be repaired. The sand which has found its way inside must be removed, the wells deepened and their timbers reinforced. This work generally takes place in November-December.

The dipper consists of a pocket made from goatskin, or better still sheepskin, hanging by means of twisted bark from a circle of wood, itself attached to a rope /keti/ of Acacia seyal bark. More often than not, it is the women who draw water; but if a man is present, he gives a helping hand: he may take his turn at drawing, or help to hoist the dipper up to the mouth of the well. Water is poured into drinking-troughs, /bátî/, hollowed out of tree trunks, generally Acacia Faidherbia /teli/, occasionally Commiphora /tógórò/; but this is rarer. It is common for the trough to bear the carved mark of the owner's clan.

The women or men who go to the well do not only water their animals; they also fill the waterskins /órî/ which carry the family supply. These /órî/ are made from uncut and tanned sheepskins or goatskins, in which the openings corresponding to the legs have been tied. They are filled through the neck opening in two ways, either directly, by using the dipper (this requires two persons, one of them pouring the water with the dipper's rope between his teeth and the dipper itself held with both hands, the other holding open the waterskin), or else by taking water from a trough with a calabash. Both techniques lead to considerable waste, as the water misses the opening, runs down the sides of the waterskin and increases the layer of mud surrounding the well. One wonders whether in the dry season people, with a little more care, could not cut down such waste: though the Zaghawa are not acquainted with any receptacle possessing a spout or lip for pouring, they often make a kind of pouring lip by folding the side of the dipper. The waterskins filled almost to bursting point are carried legs uppermost on the back of donkeys, one on each side of the animal.

People also take advantage of their presence at the wells to wash themselves and their clothes. One often sees girls holding up their wet loincloth — a large rectangular piece of indigo blue cloth — to dry in the wind, as they walk behind their herds after watering. The well is naturally also a place where news is exchanged.

LIVESTOCK

In the region of steppes which constitutes the Sudan as a geographic entity, are to be found camel breeders (called 'abbala', 'ibala' or 'albala' by the Arabs) and cattle breeders ('baggara'). The Zaghawa, as it happens, are in the remarkable position of being at the same time camel and cattle breeders.[50] The sheep are kept with the camels, travelling in large groups led by young herdsmen, whereas cows and goats do not venture far from the villages, fields and permanent wells. This dichotomy allows those men and women remaining behind to sow the fields with millet, when the rains have been sufficient.

It is extremely difficult, if not impossible, to obtain any figures on the size and composition of flocks and herds. The few facts we can adduce for the Chad are based on limited firsthand observations together with the figures obtained from government surveys (see Tables 1 and 2); in the Sudan we carried out a number of interviews in order to ascertain the ideas of the Zaghawa on wealth and poverty in their own society (see Table 4). The results show the impossibility of comparing information obtained from either side of the border: the official figures from the Chad quite clearly underestimate the actual position, whereas those of the Sudan only show upper and lower limits. Observations made at the wells during the period when the animals are brought have allowed us to make certain corrections.

If the figures obtained in the four Zaghawa districts ("Cantons") of the Chad are examined, one observes that, in 1957, for a population

Table 1. The livestock of the Zaghawa of the Chad according to several censuses

Canton	Year	Population	Taxable	Cattle	Sheep & goats	Camels	Horses	Donkeys
Kobé[a]	1953 1955	14 492	6 872	10 748	21 910	2 128	788[b]	1 097
Kabka	1953	4 136	1 836	3 085	11 550	738	164	389
	1955	4 194	1 961	3 471	8 175[c]	751	190	341
	1957	4 505	1 824[e]	4 209	8 890	784[d]	203	299
Dirong	1953	3 868	1 747	3 863	10 090	562	133	333
	1955	3 981	1 757	3 925	10 347	590	148	355
	1957	4 160	1 544[e]	4 025	6 445[c]	676	134	195
Guruf	1953	1 702	783	1 401	4 395	186	80	338
	1955	1 833	831	2 061	3 180[c]	250	81	339
	1957	1 978	747[e]	1 954	3 220	280	81	272
Total for 1955		24 500	11 421	20 205	43 612	3 719	1 207	2 132

a. In the Kobé there was only one census available, carried out between 1953 and 1955. We have been unable to obtain that of 1957, which was in progress during our stay.

b. The highest number of horses as compared with other districts is accounted for by the large number of princes, /abbo/, residing in the Kobé; at Hiri-ba their quarter alone numbered 105 horses.

c. The decrease in the number of sheep in the Kabka and the Guruf in 1955, in the Dirong in 1957, is doubtless a consequence of the great drought of 1953.

d. Consisting of 499 males and 285 females. It is obviously more advantageous to have pack camels than breeding camels.

e. The number of taxable persons decreases whereas the total population increases, presumably due to the introduction of a new category of people exempted from tax which appears on official records from 1957 onwards: namely that of mothers with three children.

of approximately 25,135 persons, there is a livestock population of 20,936 cattle, 40,465 sheep and goats, 3,868 camels, 1,206 horses and 1,863 donkeys, in other words for an average family of five or six persons about five cattle, ten or so sheep and goats, one camel and a horse or donkey. These figures however seem to be far inferior to the true position, for compared with the standards of our Sudanese informants they correspond rather to a poor man's situation. From observations on the spot we learned that the herds and flocks more likely to be met with in the Chad consisted of 15 to 20 cows and 40 or so sheep and goats. Thus the survey figures in respect of the livestock should be quadrupled at least.[51] Nevertheless these figures provide us with interesting information on the relative importance of each kind of animal from which one may conclude that the Zaghawa are chiefly breeders of cattle, sheep and goats; next come camels, whose usefulness lies in the fact that they provide good pack animals for long trade journeys. However this picture needs to be adjusted for each individual district. For from east to west, in other words from Kobé to Guruf, the number of cattle increases compared with the number of camels, as one might well have expected.

Table 2. Livestock in 1968 and 1969 (Chad)

	1968		1969		
	Area A	Area B	Area A	Area C	Area D
Cattle	65,000	450,000	80,000	20,000	1,000,000
Camels	30,000	100,000			
Sheep and goats	160,000	450,000			
Horses	8,000	25,000			
Donkeys	15,000	60,000			
Pigs	30	—			

Area A: "Sous-préfecture" of Hiri-ba, including the four Zaghawa "Cantons"
Area B: "Préfecture" of Biltine of which Hiri-ba forms part
Area C: "Préfecture" of Borkou-Ennedi-Tibesti
Area D: Eastern Chad (Wadday, Biltine, Borkou-Ennedi-Tibesti)

For the entire Chad the annual increase in livestock was of the order of 1.5 to 2 per cent.

The above information is based on the number of animals vaccinated. It was kindly communicated to us by the head of the Eastern District for Animal Husbandry of the Chad.

In the Sudan the position is distinctly different; the Zaghawa own full-size herds of camels, (certain groups that live on the edge of the desert, like the people of Anka and Musbat, in fact breed more camels than cows). This is so because the geographical location of the Sudanese Zaghawa allows them seasonal movements that are sufficiently extensive to enable them to support a large number of animals.[52] As one goes westwards, herds of cows gradually regain first place. Figures obtained from official sources in Kutum will be found in Table 3.

Nevertheless it seems that the Zaghawa as a whole may be primarily regarded as breeders of cattle and secondarily of sheep, goats, camels, horses and donkeys. One need only consider their social institutions to observe the importance of cows. Bridewealth is always reckoned in terms of cows, even if other animals are actually used for part of it; likewise, all fines are reckoned in terms of cows.

No statistics on fertility or mortality (if they exist) were available to enable us to study the growth of herds.[53] However, the French and British authorities have fought vigorously against epizootic diseases:

rinderpest, foot and mouth disease, etc. by systematically vaccinating the animals around the wells. This campaign continues to be actively pursued in the Sudan, as we were able to see for ourselves. Several of the veterinary stockmen we met were themselves Zaghawa.

Cows and goats

The Zaghawa of the Sudan only raise one breed of cows (zebus), originating from Kordofan, hence its name 'kordofaniya' or 'kordofale'. These are sturdy cows as well as good milkers. All adult cows have a name given to them before their first calf (for practical reasons which will be seen later, page 73).

The Zaghawa possess a very rich vocabulary for individualizing each animal, which may be described and named according to the colour of its coat or the shape of its horns: for example /bìrì/ "light brown", /hámárêi/ "reddish", /hárì/ "dappled", /kòrfú/, when the coat has the same colour as the fruit of that name, /mìs/ "cat", when the cow is black like local cats, or /jina/ which designates a black cow with a white tail or else white legs (or inversely a white cow with a black tail or black legs). Sometimes it is a particular feature of the horns that has drawn attention: they distinguish animals with horns pointing forwards, upwards, long or short horns, horns sloping downwards and back, one horn pointing upwards and the other downwards, and curved horns turned down along both sides of the head /kírì/. The shape of the horns may also be associated with the colour of the coat. Thus the favourite cow is a white one with short horns. Beige and red coats come next in order of preference. Again the animal's place of origin may sometimes account for its name: 'atia' "the one from Ati" (a region famous for providing good milkers). Some bulls also have a name; but this is rare and not essential, since only a small number of them, which are intended for sale and taken to market, are kept in the herd.

The goats are short-haired, with either a white coat /bei-ter/ or a bicoloured one — white and black or white and brown, in two almost equally divided patches that might almost have been marked out with a ruler. They have also a black, long-haired breed of goats, called 'hasaniya', which is again found in Libya and among the Kababish. In actual fact the animals' hair usually appears short, because it is continually cut for the manufacture of rope and carpets. These goats are bigger, provide better meat and produce more milk. Just as with the cows, names are given to goats and billy-goats although this is not a consistent practice.

Table 3. The livestock of the Zaghawa of the Sudan paying taxes[a]

Dar	Cattle[b]		Sheep		Goats	
	1965	1970	1965	1970	1965	1970
Kobé	4,120	5,474	2,670	3,548	90	148
Gala	14,868	12,676	29,960	40,161	0	0
Tuer	9,805	7,806	40,523	37,588	3,807	14,143
Artaj	3,037	3,278	4,785	5,344	5,079	7,339
Total	31,830	29,234	77,938	86,641	8,976	21,630

	Camels		Horses[c]		Donkeys	
	1965	1970	1965	1970	1965	1970
Kobé	720	842	671	762	120	480
Gala	1,127	1,414	332	226	2,204	2,544
Tuer	3,674	4,471	93	54	2,654	2,724
Artaj	1,022	1,262	82	34	1,239	1,404
Total	6,543	7,989	1,178	1,076	6,217	7,152

a. These figures do not include animals belonging to Zaghawa but recorded in the censuses of other dars which are populated not only by Zaghawa but also by Fur and Tunjur, like dar Sueni and dar Beiri. The Zaghawa population of the Dar-Fur appears from official sources (Kutum, 1970) to be around 255,000.

b. When one compares the figures for 1965 and 1970, one perceives an important reduction in the number of cattle in the dars Gala and Tuer, accompanied by an increase in the number of goats in the dar Tuer and sheep (and perhaps goats?) in the dar Gala. For the situation is not very clear in the latter country where sheep and goats appear to have been recorded together. It may be that, due to the drought, there has been some alteration in the composition of the herds and flocks. This change is not to be found in the dars Kobé and Artaj which show a fairly substantial increase in every kind of animal (to the exclusion of horses) over the five years in question. These two dars may have been less affected than the rest by the years of drought.

c. As was observed in the Chad, horses are very highly valued among the Kobé. Though few in number in the Sudan, the Kobé possess over 70 per cent of the horses. Furthermore, while the number of horses has gradually decreased in every dar between 1965 and 1970, dar Kobé alone has seen its number go up in spite of the years of drought and of the horses' special needs.

Camels and sheep

The camels (they are in fact dromedaries) also belong to a race that came from the east, from the country of the Beja. The word for them is 'bišari'. The Zaghawa prize most of all the white camels, which they consider to be pure-blooded, and next the black ones, whose exceptional resistance to thirst is valued and which are preferred as mounts for long journeys. Grey camels are reserved for short trips. The smaller Bideyat camels are not highly thought of.[54]

Besides providing their camels with grass and water, the camel-breeders see that they receive a supplement of natron (sodium carbonate). This is periodically given to them and, when it is very hot, even once or twice a month, after it has been prepared by being ground, mixed with water and sometimes with cow dung. The amounts given annually vary between two and four camel-loads per herd of about twenty animals; since a camel-load weighs approximately 130 kilos, this means that each camel requires between 15 and 20 kilos annually.[55]

The Zaghawa raise two breeds of sheep: black, long-haired, short-tailed sheep found in the regions of Kornoy and Tiné as well as the Chad, and the so-called "red" short-haired sheep, which is sometimes white or brown with white spots, with a big, thick, fat tail, which are particularly found in the regions of Musbat and Forawiya. The black breed is called /gèrèj/. The people of Musbat dislike it, for it gives less milk and its meat is not as good. But it is a sturdy breed, which stands up to the cold and contents itself with poor grass; it can live in a rocky or mountainous region, whereas the other breed could not do so. Both males and females of the /gèrèj/ breed bear horns, much larger in the case of the male than the female.[56]

The other breed is called /kābāš/, because it was originally obtained from the Kababish; it is also known as /seri/. It yields more milk and its meat is excellent. It is distinguished by a huge tail, which can provide a meal for three, a dewlap just like those of cattle and a very large girth. The male is bigger than the female; neither have horns. The Zaghawa try to have two or three /kābāš/ rams in each flock for selective interbreeding. It is these "red" sheep that are taken every year in flocks from Forawiya and Musbat to Khartum to be sold, especially at the time of the moslem festivals.

Horses and Donkeys

Horses are the chiefs' mounts and their food is the object of special care. In addition to grass, during the whole year round or else simply in the rainy season, they are given each day a certain amount of milk, just over two litres at a rough estimate. This corresponds to the capacity of the average calabash or the wooden dish /godu/ in which millet porridge receives its shape or that of a small trough /madi/. Thus throughout the year the horse of 'mogdum' Fodul of Oru-ba consumes the milk of two cows. It also receives 'kreb' or millet in the form of grain, balls of flour kneaded with water or milk, or else as porridge; a field /hirde tibe/ "the horse's field" is specially cultivated for it. Women prepare leaves of the /núrdà/ tree for the horses. These leaves are kept in granaries and boiled in water when required. Women also give the horses 'askanit' (Cenchrus biflorus) seeds, which are stored after the rains and then crushed to remove the thorns. When a mare dies and leaves a foal, the latter is fed on cow's or goat's milk.[57]

Donkeys, though ridden by women, blacksmiths and the very poor, serve mainly as pack-animals to which all the chores, like carrying water, wood or taking goods to market, are assigned. They are not treated with any special care. In the Sudan there also exists a breed of tall donkeys, the 'rifai', which are highly valued.[58] They are the favourite mount of the 'faki', who do not ride horses.

RIGHTS OF OWNERSHIP AND USE

Wells

Strictly speaking wells have no particular owner. When asked about this, most people reply: "a well belongs to the one who dug it, but he may permit others to draw from it". This assertion is correct; however the phrase "the one who dug it" must be understood as referring not to an individual but to a group, such as the inhabitants of the same temporary hamlet ('ferik'), of the same village or the members of the same clan. But if one of the group has refused to take part in the operation, he will not be able to gain access to the well later on. It is therefore necessary to have participated in its construction in order to be one of the legitimate beneficiaries. The permanently constructed wells, built by the government, are open to all. Only the well dippers are privately owned. They are never left by the wells but are brought there on each visit.

In practice few clashes occur when there is enough water and the animals are in a healthy condition, even if someone outside the group comes to draw water, provided he behaves according to the established custom, by bringing his own dipper, waiting for his turn, and so on. But should water become scarce or a man try to water sick animals, daggers are then drawn to settle the matter.

For instance in October 1966, Adam Kitir, an uncle of our young friend Zakaria Fodul, met his death in the following way. A certain number of herds of the Oru-ba region had contracted rinderpest. A few herds however were spared, among them that of Adam Kitir, a herd of about fifty cows, seven or eight camels and approximately sixty sheep, and the lucky owners decided to keep a watch on the wells to prevent their animals from coming into contact with contaminated or suspect animals. The owner of an infected herd took this amiss and a fight followed in which five people were killed including Adam Kitir.

Around the wells are to be found a certain number of wooden troughs hollowed out of tree trunks and bearing one or more clan-brands. In the Sudan we noticed that the most frequent mark to be seen on the troughs was the clan-brand of those people who had dug the well. The man who hollows out the trough carves his mark on it (obviously he must have taken part in the construction of the well). He has a right to use the trough to water his animals, as have no doubt the members of his clan; others outside the clan may only use the trough after them.

Pastures

The pastures, 'goz', 'jizu' and grazing lands situated near the pools and wells, are common property; they are not owned by any clan, village or settlement. Anyone may freely take his herds and flocks there, provided he does not hinder or drive away those already settled there. Thus the only right is that of first arrival. As a rule the herdsmen take their flocks and herds to the same spot each year; but they find out beforehand about the state of the pastures, the supply of water and the presence of groups that have already settled in the region.

At Forawiya two or three men set off in advance on their camels on the look-out for good pastures. If they find any, their group will make its way there with the herd and flock. But if they come upon other herdsmen already there, they continue their search elsewhere.

In the region of Kamo, there is a pool called /dugu-ray/ ("to think up an idea") where the inhabitants of the neighbouring villages meet to hold a council at the time of /tàrbà/ — in other words after the rainy season, towards October — in order to choose a place well supplied with water where they will establish their camp (Ar. 'ferik'). In this period the pool is already dry.

The pastures we mention are frequented not only by the Zaghawa, but also in the eastern part by Arab tribes like the Kababish and the Ziyadiyeh, and in the western part by a few Bideyat and Daza. In theory all should go smoothly: "if a flock or herd already occupies the place where you wish to go, you are to move on and settle further away, at a sufficient distance to prevent your animals mixing with those of your neighbour". In practice there are almost daily fights for the pastures and the watering points, mostly in the dry season.

Flocks and herds

The same flock or herd may include animals belonging to different people such as the head of the family, his wives or mother (who own animals in their own right), his children (who receive a few head of cattle as soon as they are circumcised) and lastly certain relatives, especially brothers. As a rule members of the same family living together in the same compound or else in proximity group their animals for grazing.

Animals owned by the same person may have been acquired in quite different ways. A number have been received as bridewealth on the marriage of daughters or close female relatives, a few may represent the payment of a fine following some misdemeanour, and some have been inherited. Occasionally they are bought. In a few cases they may represent a share in the spoils taken in the course of a raiding party among neighbours or constitute the product of a theft by the man on his own. Only camels are stolen.[59]

Each owner marks his camels with the brand /erfe/ of the clan to which he belongs by paternal filiation. The animals are branded with a red-hot iron, usually on the cheek, neck or thigh (either the right or left one). The mark may consist of one or several clan-brands. Sometimes the camel also bears the brand of the owner's maternal clan — namely that of his maternal uncles and their sons — when he prides himself on belonging to it. The owner for similar reasons may also add the brand of his wife's clan, in other words that of his father-in-law and brothers-in-law.

Each brand is given a name. Sometimes it is the name of some common object: /seli/ "sword" is the brand of the /aŋu/ clan; /mamur/ "copper bow", the brand of the /genigergeʀa/ and the /imogu/; /ágli/ "hooked stick" (for knocking down the fruits and young leaves given to goats), the brand of the /elbira/ (to give just a few examples). The brand may also represent some familiar shape like /gwa dei/ "crow's foot", the mark of the /ohurra/.[60]

It is important to know that each clan has its brand, that certain related clans use the same brand or variants. Every camel owner marks his animals with his clan-brand, as a proof of ownership in case the animal is stolen or gone astray. It sometimes happens that a thief tries to confuse the mark by modifying the brands or superimposing his own clan-brand; but as the camel normally bears several brands representing the previous marriages of his owners, it is difficult to fake the mark in this way.

It is these clan-brands that are often found carved on the sides of troughs, on stones near the wells and on young herdsmen's wooden weapons. Some are also to be seen drawn in charcoal inside rocky shelters, alongside ancient cave paintings.

The loaning of animals

It is fairly common practice for an owner to lend a few animals to a less well-off relative, friend or fellow villager. Cows, camels and goats are lent in this way. These loans are not intended to allow the borrower to set up his own herd or flock with the product of the animals lent, as has been seen with other peoples; goats and cows are supposed to provide him with milk for his own use, and camels are lent for transport.

The Zaghawa are quite prepared to lend their animals to any man who takes good care of them, but in the opposite case, they will refuse. An animal is usually lent for a year or two, but this can greatly vary. The borrower will only take the milk; the offspring (young camels, calves and kids) will revert to the owner. If the loaned animal dies, the borrower is not bound to replace it, but he must prove that the animal is well and truly dead by showing the owner its skin, ear or tail, or else by producing witnesses. Certain informants disagree: if the dead animal, they say, belongs to a member of the borrower's family, it is not replaced; but the owner is entitled to compensation if he is a stranger to the borrower.

It is common practice to borrow a camel from one's neighbour, whether relative or friend, for transport or a long journey. A man who wishes to go to the salt-mines of Démi to fetch rock-salt will borrow two camels, in order to bring back one load of salt for their owner and another for himself. According to some informants, should the animal meet an accident on the way, the owner is not entitled to claim when the transport has been partly undertaken on his own behalf; but he is entitled to compensation when the transport has been carried out on the borrower's behalf exclusively. One may also lend one's horse. The same term: /tow/ is used for lending and borrowing.

Such a perfect display of community spirit in theory does not prevent actual clashes in real life. It is common to see people up before the courts claiming animals lent long ago, which have not been returned to them and of which the borrower declares himself the true owner. The animal in dispute may sometimes have even been disposed of and therefore no longer be in the borrower's possession. Occasionally the latter, unable to return the borrowed animal, proposes by way of compensation a donkey in place of a cow simply because it is his only possession. During our inquiries we came across a few cases where the lender of a cow receives another animal as surety from the borrower, but this does not seem to be the general practice.

Thus borrowing animals will not enable a poor man to set up his own herd or flock. He may solicit gifts from rich relatives, occasionally from the chief of the district or even from the sultan, but

ultimately the most effective way open to him is to emigrate and take
a job as a paid worker on building sites in the Sudan, or on the cotton
plantations of the Gezira, or, not to go so far away, hire himself
either as a labourer in the millet fields in the El Fasher area or
eventually as a herdsman to a wealthy cattle owner with a view to
saving up some money and buying a cow.

SEASONAL MOVEMENTS OF FLOCKS AND HERDS

From the outset of our first stay among the Zaghawa of the Dar-Fur
during November-December 1965, in the period called /dâbó/ (Ar.
'šitā'), we immediately noted that the seasonal movement of flocks
and herds was a subject that frequently and spontaneously cropped up
in conversation. In Anka at the beginning of November we heard it
said that the flocks and herds were then in the north at the pool of
Jinnik, where a fight between Zaghawa and Arabs had just taken place;
at Um Buru the people spoke about their flocks and herds that had
gone to the 'jizu' (Z. /séʀí/) far in the north; at Tiné in December
the sultan Daosa was concerned about his camels, deep in the south
at the time in the Dar Masalit under the care of an Eregat herdsman,
and so on. Furthermore an excellent article by Talal Asad on the
seasonal movements of the Kababish Arabs of North Kordofan, the
far eastern neighbours of the Zaghawa, which we read at this time,
prompted us to seize the opportunity of making a systematic inquiry
into the movements of animals in the different Zaghawa groups of
the Dar-Fur: Artaj, Tuer, Gala, Kobé, various clans grouped around
Anka and Dor, and Kabka from the region of Tundubay.[61]

We had not previously conducted an inquiry of this kind in the
Chad (although some information was later obtained from Zakaria
Fodul, who came from Oru-ba); nevertheless we do not think, although this would obviously need checking, that the Zaghawa of the
Chad practise pastoral transhumance on such a large scale as those
of the Sudan. We see two reasons for this, the second of which
(based on the composition of flocks and herds) follows from the
first. In the Sudan, Zaghawa country is not closed in by any natural
barrier in the north and opens onto the Sahara, offering the flocks
and herds, in the rainy season, pastures where the animals can
travel and feed for several days — these are the famous pastures
known as 'jizu'; in the Chad, however, Zaghawa country is bound in the
north by the range of the Ennedi, where the Bideyat live. The Zaghawa
entrust the latter, to whom they are related, with their flocks and herds
to be taken to the pastures of the Sahara. However the Bideyat are

above all concerned with feeding their own flocks and herds, and those of the Zaghawa only take second place in their preoccupations.

As a result the Zaghawa of the Chad have far fewer camels than those of the Sudan; speaking in more general terms, one might say that camel herds decrease in size from east to west. For whereas in Anka one may find owners of 150 camels, in the Guruf a camel owner has at most two or three animals. A more detailed picture would naturally take into account both the relative importance of agriculture in certain regions (more millet is grown in the Guruf than at Anka) and the composition of flocks and herds (for instance at Tundubay, where hardly any camels are to be found, there are large herds of cows).

One of the questions we regularly asked our informants may cast some light on this second point: what is the composition of the herd of a man regarded as rich by the inhabitants of the village, and what is that of a poor man? The second question always seemed to cause a great deal of embarrassment and was usually left unanswered; on the other hand our informant or informants would answer the first confidently and without hesitation (cf. Table 4).

Anka is certainly the place where the largest number of camels are owned in Zaghawa country. Here a rich man's herd consists of about 150 animals. As previously mentioned, the size of camel herds constantly decreases towards the west: thus among the Kobé of the Sudan a rich man possesses about fifty camels. These figures seem much higher than those we noted in the Chad.[62] For sheep it is the opposite: at Anka a rich man would own 200, in the region of Dor, as in the dar Tuer and the dar Gala, 300, and in the Kobé 400.

The information collected in respect of cows shows herds of almost equal size in the different regions concerned. The number of animals varies around 100; it goes down to 50 in the region of Dor.

As regards goats we often heard it said that "a rich man doesn't own goats"; only the Gala and Kobé informants gave us some information: the former mentioned a herd of 400 to 500 goats, the latter 100 goats. A poor man's herd is as little as 10 or 15 or even 5 to 10 head. Nevertheless there is no one, however poor, who does not own a couple of goats.

A rich man owns horses: 15 to 20 at Anka and among the Kobé, fewer in the other regions (4 or 5 for the Gala; only one or two at Tundubay, for fear of the lions and hyenas that abound there); he also owns donkeys: 10, occasionally 20. A poor man evidently does not have a horse but he may own one donkey.

As for blacksmiths, who once were only permitted to own a few goats and donkeys, they now own sheep and occasionally a few cows, though this is rare.

Table 4. Notional size and composition of flocks and herds among the Zaghawa of the Sudan[a]

	Camels[b]		Sheep		Cows	
Owner:	rich	poor	rich	poor	rich	poor
Anka	150	15	200		100	7/10
Dor	60/70		300		50	
Tuer	60/70		300		50/60	
Gala	50	1	300	0	100	5
Kobé	50		400		100	
Tundubay			150		70/80	1

	Goats[c]		Horses		Donkeys	
Owner:	rich	poor	rich	poor	rich	poor
Anka			15/20	0	10/15	1/2
Dor						
Tuer			4/5		10/12	
Gala	400/500	10/15	4/5	0	10	1
Kobé	100		20		20	
Tundubay		5/10	1/2			1

a. This Table presents the results of the inquiry described and commented page 50. It does not refer to any census but to the opinion the Zaghawa form of wealth and poverty in every district. As for the Chad we collected only two answers: At Oru-ba (in the Chad Kobé country) a rich man owns approximately 50 camels, 300 cows, 100 to 200 sheep and goats; a Guruf informant mentions 15 cows for a small herd and 50 for a large one.
b. There are no camels at Tundubay.
c. At Tundubay a rich man has no goats but sheep instead.

Seasonal movements of flocks and herds at Anka

Among the Zaghawa, as we have seen, the people of Anka are the largest owners of camels in a region where less grass seems to be available than elsewhere. This is because their country has no northern border. One need only climb the hill, a heap of rocks, that overlooks the wells and village of Anka to catch sight of the Sahara that extends as far as the eye can see and which the rains will transform into a vast pasture.

The first falls of rain are the signal for the herds of camels and flocks of sheep to depart for the vast sandy expanses of the 'goz' (Z. /šige/). At this time of year, /gyé/, they will find grass and water already there and it is quite easy to move about, as there are no wadis to cross. Flocks and herds make for /Yumin šige/ or /Bui/, in the 'goz' el Harr, el Arab or Leban. They spend the whole of the rainy season there under the care of young herdsmen.

During the /tàrbà/ season (October), they disperse towards the large pools of Um Shedar, Tari-mara and Jinnik, in whose proximity they remain till the end of December. There the flocks of sheep and herds of camels meet up again with the herds of cows; these do not go onto the 'goz' during the rains but stay by the villages.[63]

With the end of /dâbó/ and the beginning of /âigì/, the dry and hot season, the flocks and herds draw gradually closer to the permanent wells (including, among others, those of Anka) and the herdsmen set up their camps /guli/, (Ar. 'ferik') in the wadis around the wells.

Seasonal movements of flocks and herds at Dor

Immediately after the first rains, as at Anka, the flocks of sheep and herds of camels make for the 'goz'. They go to Gumgum, in the goz el Harr, Nay, Koru-hay, Kharra and Khurr, and to jebel Jak. They move around the goz until October.

During /tàrbà/ they leave the 'goz', which do not retain water; some make for jebel Kherban and other for jebel Imam. They circulate between these two points, where there are vast expanses of dry grass which is very good for the animals. The flocks and herds go and drink at the nearest wells of Bir tomur, Bir iz el-khadim or Bir fokhma; the sheep are watered every other day, the camels only once a fortnight which seems rather surprising but was confirmed by the informants concerned. The animals remain there till the first rains, when they make once more for the 'goz'. A few flocks and herds from the Dor region may go beyond jebel Imam during /dâbó/, heading for the wadi Tega.

Compared to the circuit of the animals at Anka, it is a simpler one that amounts to a to-and-fro movement between the 'goz' and the pastures of the dry season, with water obtained at the nearest well. This is nothing like the complete dispersal described at Anka which is made possible by the presence of large temporary pools. Still the area accessible to the people of Dor includes numerous 'goz' but few large pools; there is in fact a large one at Jinnik but they do not seem inclined to frequent it, perhaps because of its being already quite heavily crowded by the Zaghawa of Anka besides a few small Bideyat and Arab groups. It is true that the people of Dor, the southernmost Zaghawa, own far fewer camels, only half as many as those of Anka.

In the rainy season their cows, which are also less numerous, remain near the villages and during the other seasons they stay in the 'ferik' close to the wells, especially around Dor. The goats are kept near the villages or go with the cows.

Seasonal movements of flocks and herds of the dar Artaj

The movements of camels and sheep are more extensive in this region than at Anka and Dor. During the rainy season, /gyé/, the flocks and herds move in the area comprised by Um Haraz and the wadi Ambar, where they find good pastures, thus avoiding the clay soils and their dampness.

At the onset of October they go off to the north and the northeast onto the grazing grounds known as 'jizu' (Z. /sérí/). They first make for the furthest pastures which they use during the short period that corresponds with /tằrbà/; at this point they are several days distant by camel from Um Haraz, beyond Mahallat kwoila in the direction of the 'goz' Niuw.[64] Occasionally the flocks and herds go as far as Bir Atrun (the Zaghawa's /teɒi-bà/). This is the only moment in the year when these desert pastures are usable, before they become scorched by the sun. It is worth noting that the flocks and herds are driven to the furthest point in two groups, leaving between them an ungrazed area which will serve as a corridor on their way back. This technique is common practice whenever flocks and herds go off to distant pastures.

During /dấbó/, that is to say from November to January and even in some years until March-April, the flocks and herds continue to graze on the 'jizu', but they abandon the furthest points to graze within areas to the north and mainly east of Um Haraz at five or six days distance by camel. They will now be moving near Attawia, Misallakhat, Tuwanis, Rakib and in the region around Ha Wiša (places not located on the maps). The animals do not drink at the wells since they get enough water from the fresh grass; the herdsmen accompanying them live off milk.

Starting as early as February and sometimes much later, depending on the year, the flocks and herds draw closer to the permanent wells and to the pastures near those wells; the wells of the wadi Fokhma are the farthest in the east and see less animals than those of Musbat and Orori, north of Um Haraz. There, they meet up with other flocks and herds belonging to related clans or to neighbouring tribes. One finds assembled there people from the dar Artaj, and a large number of Ila Digen and BiriaRa, also some Bideyat from the Ennedi, as well as Ziyadiyeh and Kababish Arabs from the East.

Cows and goats spend the whole year in the pastures close to the wells of Um Haraz, Orori and Um Marahik.

Seasonal movements of flocks and herds in the dar Tuer
(region of Musbat)

Here the picture is very much the same as in the neighbouring dar Artaj. During the rainy season, camels and sheep go onto the 'goz', east and northwest of Musbat, around the wadi Maghreb; the goats are sent out together with the camels and sheep, especially in the Musbat region. This is the first stage of the migration northwards. The animals spend about three months on the 'goz' drinking from the pools.

If there is much grass, they may stay longer. But usually once the rains are over they start on the second stage, heading north to the 'jizu' pastures. They go to the north and northeast of Sendi and Tekwoila. Like the flocks and herds of the dar Artaj, they begin by grazing on grass which will be scorched first by the sun; in some years this course may lead them as far as Bir Tundub, north of latitude $18°$. They remain in the 'jizu' for five to six months, and then they go south onto the pastures three or five days by foot from Musbat.

During the dry and hot season, that is to say from February, or at the latest from March-April, to June, they return towards the permanent wells and wadis; there the herdsmen set up their 'ferik'. The grazing grounds are around Musbat, Forawiya, Shiget karo (or Shigeit-karo), Ba-meshi, in the wadi Lil, at Turba, Bir En and Arkuri.

Seasonal movements of flocks and herds in the dar Gala

We collected information in two places sufficiently apart to provide significant differences: Kamo, east of Kornoy and at the same latitude ($15°$), which is in a region where wadis and wells are numerous; and Forawiya ($15°$ 20') in a much more arid area at the edge of the desert.

The people of Kamo and Darma, slightly further to the north, have, so to speak, a great deal of grass, much more anyhow than there is at Anka, Um Haraz and Musbat and consequently they raise fewer camels and more cows. There a rich man has at most 50 camels and 100 cows: about ten times more than the average head of family (5 camels, 10 cows, 50 sheep, 10 goats, one horse and one or two donkeys). Of course the movements of the animals are more limited in range.

During the rainy season the flocks and herds of Kamo go into the region of Dugu-ray, on the border between the Sudan and the Chad

and those of Darma go towards the wadi Gardai; these are areas
crossed by few wadis where as a consequence water is scarcer. At
the end of the rainy season both groups gradually make their way
towards the wadi Hawar and the pastures of the 'jizu' that lie beyond
it. But in a good year, that is to say if there have been sufficient
rains, the flocks and herds do not cross the wadi Hawar. They travel
south of this area and go east of Bow-ba, up to the wadi Forawiya
where they find water to drink. However there is no need to cross
the wadi Forawiya for new grazing grounds and they keep within Gala
territory. They may remain in this region for six to seven months,
up to January-February, stopping at the pools to drink.

Generally speaking, the wadi Hawar may be considered as the extreme limit of the movements of the flocks and herds coming from
around Kornoy, a relatively fertile region watered by a number of
wadis. The wadi Hawar is sometimes crossed, when the rains have
been scarce, in the search for more extensive pastures; though it is
more usual in such a case to decide on an earlier return southwards
of the flocks and herds. The wadi Hawar is also crossed when excessive rains have resulted in a larger number of mosquitos.

During the dry and hot season /ãlgì/ the flocks and herds find their
food in the wadis around Kamo. They graze in the wadis Kamo, Kulkul, Am Sugat, Tama-jora and Adera. They may go anywhere but
have to avoid the region of Bow-ba which is occupied at the time by
the flocks and herds of the people of Forawiya. Such a relatively
small range of dispersion of flocks and herds during the dry and hot
season indicates the presence of a greater number of permanent
watering points in the dar Gala than in the other Zaghawa dars.

The cattle of dar Gala benefit from the greater proximity of the
wadi Hawar and have only to make a short journey to reach the salt
pastures (sabka) in the bed of the wadi. Thus, as soon as the rains
are over they make for the wadi Hawar and spend October there.
This movement of the cattle is peculiar to the dars Gala and Kobé
and is permitted by the configuration of the wadi Hawar in these parts.

When this course of grass salt is over the cattle return to their
own villages and disperse among the neighbouring hills where the
camels and sheep will come and spend the dry season with them. In
this period all the animals drink at the same wells: namely those of
Kamo, Bi-mara, Kulkul, Adera, Er-shehari, Nana and Di-sheri.
During this season camels are watered once a week in the evening,
sheep every three days around midday and cows every other day in
the morning.

The animals of Forawiya, that is to say camels and sheep, do
not enjoy such good conditions as those of the Kamo region. Every
year in October, they cross the wadi Hawar immediately after the

1. The area under examination

The numbered squares correspond to sketches 2.1 to 2.5.

Sketch established after the maps of the Survey Office : Kornoy. 1931; Musbat, 1934; Umdur. 1938; Wadi Hawar. 1942; Kutum. 1952 (scale : 1 : 250.000). Though unaccurate in respect of certain features they have proved extremely useful in the main.

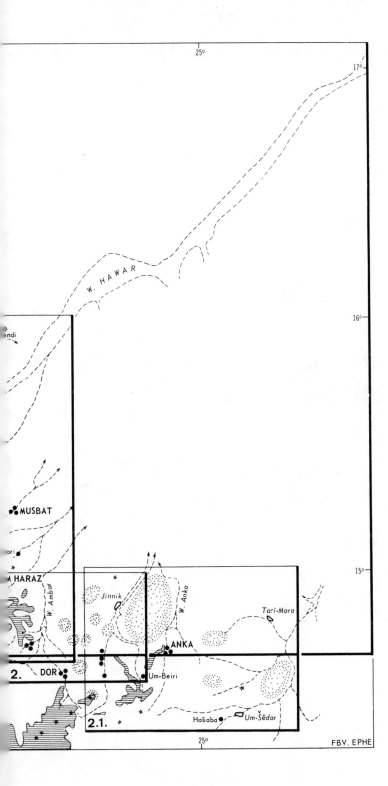

2. Seasonal movements of flocks and herds. A regional study.

SCALE : the sketches numbered 2.1 to 2.5 are on the same scale (1 : 1,000.000). in order to make it possible to appreciate the difference in range of flocks and herds movements from one region to another.
However the first two sketches (2.1 and 2.2) which depict relatively confined movements have been enlarged to become more easily read.

ARROWS : their path indicates the movements of animals; their markings represent the season during which the movements took place. <u>Black</u> is for water (movements during the rainy season, pools and wells). <u>White</u> is for drought.

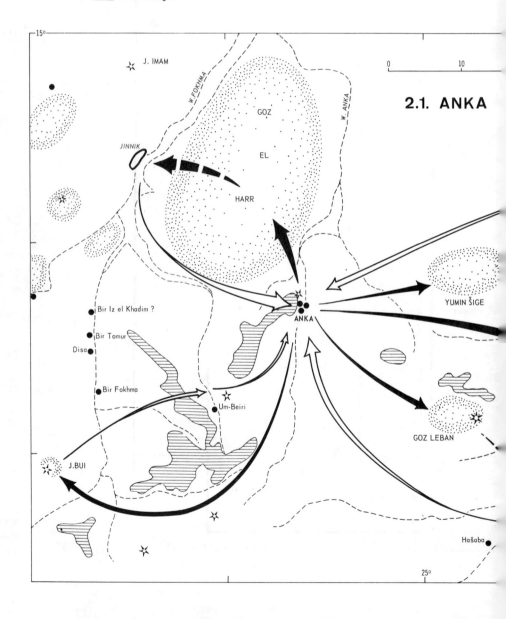

2.1. ANKA

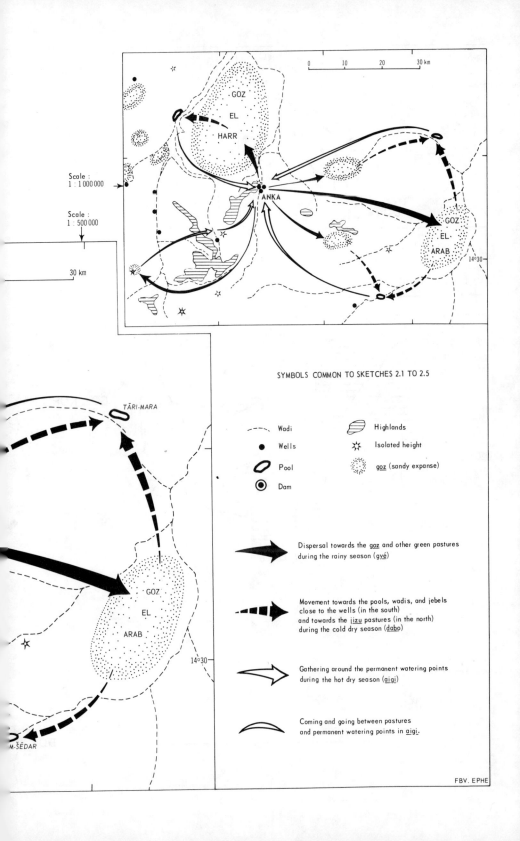

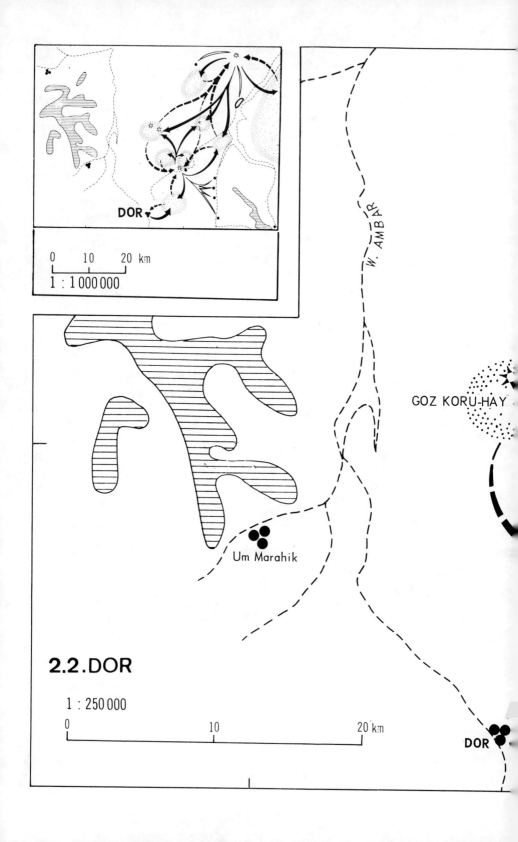

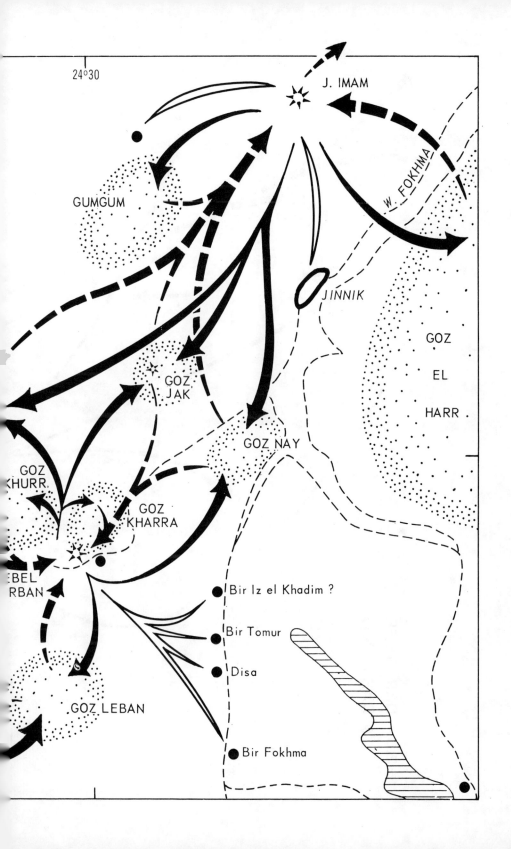

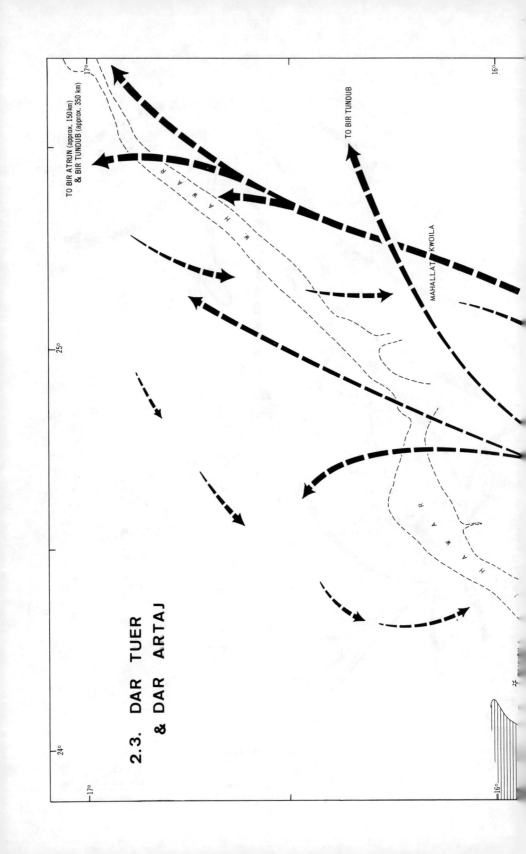

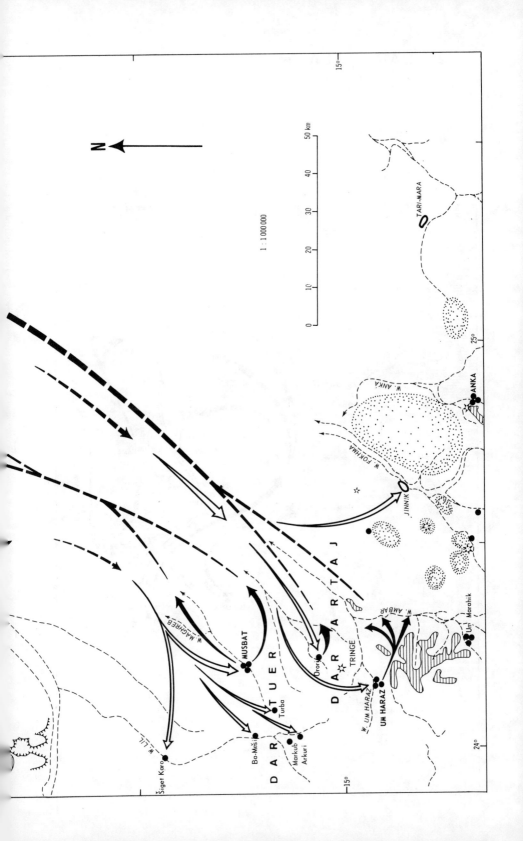

2.4. DAR GALA (KAMO)

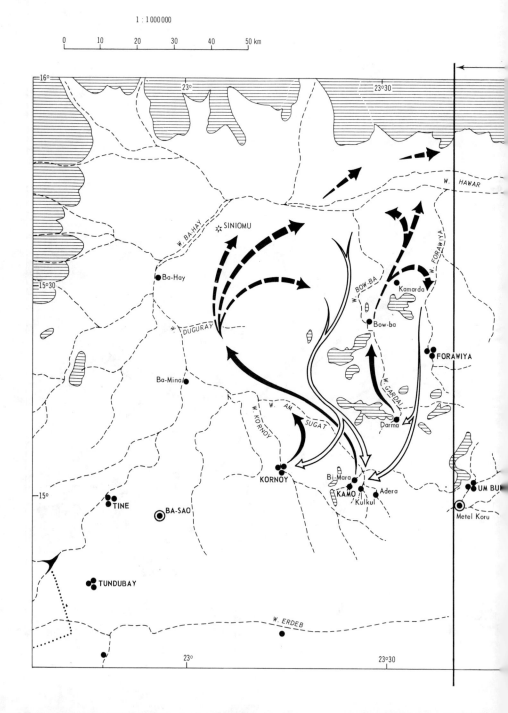

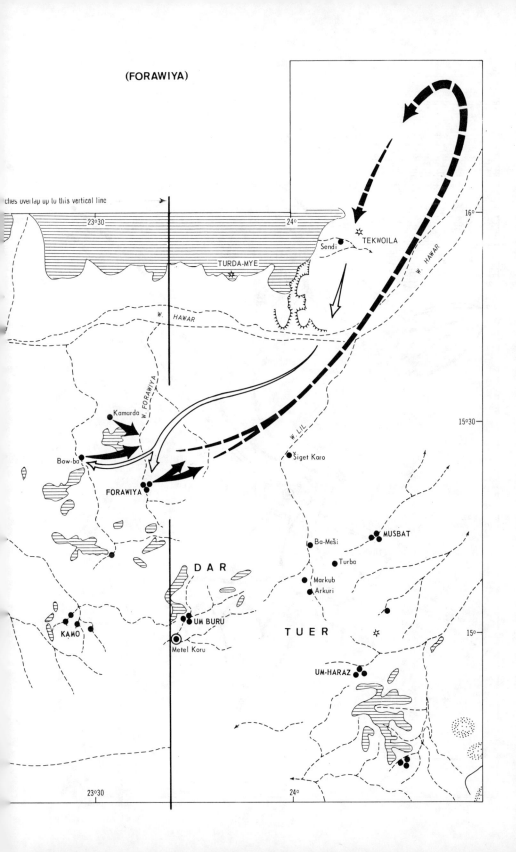

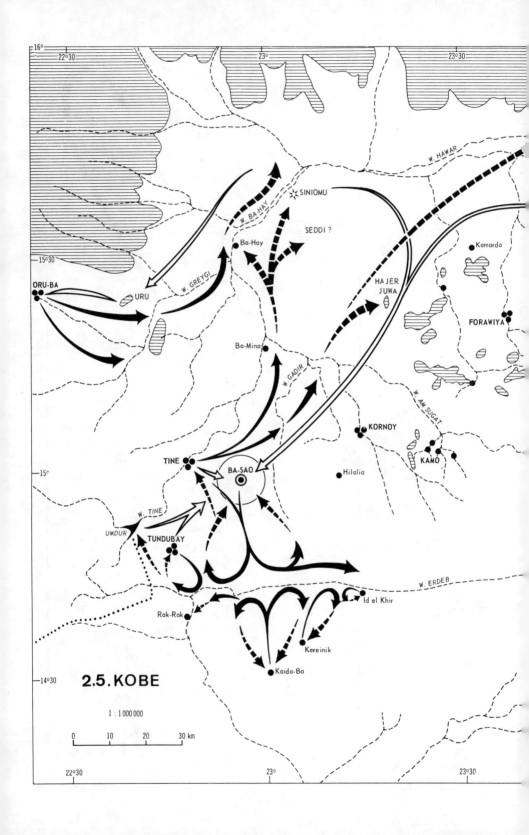

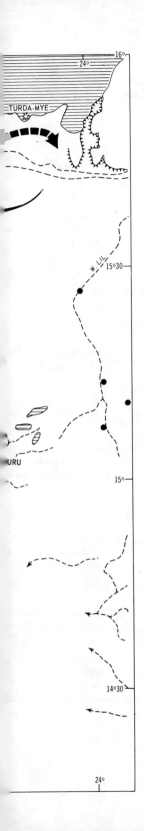

Table 5. Calendar of the movements of camels and sheep among the Zaghawa of the Sudan

```
              15 May June July  Aug. Sept. Oct. Nov. Dec. Jan. Feb. March April M
       200 mm
  rain 100
       0
            ìrsási              mùgùlí
                  - - - gyé - - -→     ← - - - dábó - - -→ - -   áìgì - - →
                                    tárbà
                                    - - - - - - - ■■■■▶
              ← - - wet and cool - - →   dry_and_   _dry_and_  _ _ _dry_and_
                                          hot          cold           hot
```

Anka	outward expansion towards the sandy expanses (goz)	maximal outward expansion towards the large temporary pools	retreat towards the permanent wells
Dor	outward expansion towards the sandy expanses (goz)	to and fro movement between the pastures of the dry season and the wells	retreat towards the permanent wells
Um Haraz	dispersal towards the wadi Ambar	maximal dispersal northeastwards (jizu)	return towards the permanent wells
Musbat	dispersal onto the sandy expanses (goz) in the region of the wadi Maghreb	maximal dispersal northeastwards (jizu)	return towards the permanent wells
Kamo	dispersal within the region comprised by Dugu-ray and the wadi Gardai	towards the wadi Hawar and beyond it towards the pastures at jizu if the rains have been insufficient	loose regrouping in the region of the permanent wells
Forawiya	dispersal in the region of the wadi Forawiya and its tributaries	outward expansion beyond the wadi Hawar	return towards the permanent wells
north of Tiné	dispersal towards Gadir and Ba-mina	outward expansion towards the jizu pastures	gathering around the wells and the Ba-s dam
south of Tiné	dispersal towards the wadi Erdeb	return towards the pastures close to the wells	gathering around the wells and the Ba-s dam

68

floods are over, that is to say when there is no fear of the wadi running again or of mosquitos being about; they are going north in search of pastures, two or three days journey away.

The people of Forawiya, the majority of them /genigergeʀa/, frequent the same 'jizu' pastures as the Kababish, but their 'ferik' are always separate. According to the 'umda'[65] of Forawiya, Zaghawa and Kababish live in harmony, though this does not prevent quarrels among herdsmen. As long as grass is plentiful, the flocks and herds remain in the 'jizu'. The return journeys accordingly vary from one year to the next and may take place between January and March. Each one then returns to his own permanent well: Kamarda, Bow-ba but mostly Forawiya.

The cows, like those of Kamo, never cross the wadi Hawar but remain in it for a month in /tàrbà/, the month following the rainy season, in order to graze on the salt pastures (sabka). Then they return to the 'ferik' near the wells: Bow-ba, Kamarda, Forawiya, Darma, Kamo, Teri-ba, Huma-ba, Dabay, Gerdi, Key-ba, Huriokoila, Sunjabak, Dildila, etc.

In a sketch of the seasonal movements of nomads in western Sudan, K.M. Barbour shows only one circuit for the Zaghawa, which takes the people of the Forawiya region to the west of El Fasher where they are supposed to stay from November to June. No Zaghawa has ever mentioned this circuit to us as a regular transhumant movement for animals. However it does correspond to a trading journey which takes the Zaghawa southwards to provide themselves with millet.[66]

Seasonal movements of flocks and herds of the dar Kobé

We are dealing here with the smaller part of the dar Kobé that belongs to the Sudan Republic, namely the region around the wells of Tiné and Ba-sao in the centre, Ba-mina and Ba-hay in the north and Kereinik and Kaida-ba in the south. The study of the seasonal movements of the animals from the dar Kobé brings to light the impact of a new geographical factor, namely the existence of the vast bed of the wadi Erdeb, crossing the southern part of the country from east to west and offering another area of attraction of a quite different type to the flocks and herds of the Kobé Zaghawa.

The pattern of their movements as described by a number of informants is as follows. The people living north of a Tiné-Kornoy line send their camels and sheep to the 'jizu' in the region of Siniomu, bordering with the Bideyat country, and beyond the wadi Hawar in the Turda-myé area. The country south of the Tiné-Kornoy line is divided in its turn in two zones by the wadi Erdeb, running from west to east in a straight line roughly parallel to the Equator. The people that live in the northern zone (between the wadi Erdeb and the Tiné-Kornoy line), such as the villagers of Tundubay and Ba-sao, travel southwards in order to reach the wadi Erdeb. To rejoin the same wadi, the people that live in the southern zone (south of wadi Erdeb), such as the villagers of Kaida-ba and Kereinik, have to make their way northwards.

The actual circuits are more complicated. At the very beginning of the rainy season, during /ı̃rsã̃si/, those people living north of the Tiné-Kornoy line make their way north to Gadir, Ba-mina, Surajari (which we have been unable to locate), while those from Tundubay and Ba-sao, going southwards, travel directly to the wadi Erdeb in order to take advantage of the early rains of this area, which have caused the new grass to appear. They remain in their respective areas for the entire duration of the rainy season, /gyé/, with the flocks and herds drinking at the pools.

During /tã̃rbã̃/ and /dã̃bó/ those that have gone northwards continue their movement: they make for Ba-hay, Seddi, hajer Juwa, then Sini-omu and Turda-myé, in other words for the 'jizu' pastures where they move around, grazing there for as long as possible. They are about five or six days distant from Tiné. The herdsmen are usually Bideyat, relatives of the owner or owners. In a good year, according to some informants, these will not return until March-April.

Those flocks and herds that have gone to the wadi Erdeb, return earlier to spend /dã̃bó/ (the dry and cold season) in the pastures close to the wells: villagers from the northern zone rejoin Tiné (the Kobé), Tundubay (the Kabka) or Rak-rak, that is on the border with the Gimir; villagers from the southern zone rejoin Kereinik or Kaida-ba. The transhumant animals drink at the same wells as the cows: every other day in the morning in the case of cows and sheep, every four or six days in the evening for camels.

During /ã̃ı̃gı̃/ all the animals gather around the wells and the reservoir of Ba-sao. Camels and sheep stay at about 8 km from the watering points because all the grass within this perimeter has been grazed by the cows that remained on the spot. Before the dam at Ba-sao was built, the watering point that drew the greatest number of flocks and herds in this season was Tiné with its wadi and wells.

To conclude, the Kobé practise two different transhumant circuits. One is similar to the movement already described regarding the dars Artaj, Tuer and Gala, that consists in a fairly long journey northwards to reach the 'jizu' pastures. The other is a movement of a type yet unmentioned, that takes a certain number of animals to the

south, at a shorter distance. This is not unconnected with the fact
that the people of Kobé own more cows than camels. As in the other
dars the constraints of the dry season lead to the regrouping of the
animals around the permanent watering places.

Seasonal movements of flocks and herds in the region of Tundubay

The region of Tundubay, though already mentioned above, deserves
a particular section. Unlike the other regions of Zaghawa country,
there are no camels at Tundubay, only cows and a few sheep. The
cows never graze very far from Tundubay. During /gyé/ and /tàrbà/
(and as late as /dábó/ if the pool of Umdur still holds any water)
they are to be found in an area comprised between the pool of Umdur,
Tundubay and the wadi Erdeb; otherwise they go in the wadi Erdeb or
to the dar Gimir, to Tiné or to the outskirts of Hilalia in the dar Gala.
During /áĩgì/ their position remains the same as in /dábó/.

Situated in the south of Zaghawa country, Tundubay enjoys a special
economic situation; the flocks and herds are far smaller and cultivation of millet is a less risky venture than elsewhere.

Seasonal movements of flocks and herds in the region of Oru-ba

Although no systematic study was carried out of the movements of
flocks and herds among the Zaghawa of the Chad,[67] we were nevertheless able to collect some fragmentary information regarding the
movements of the animals of the Oru-ba region (Kobé).

During the rainy season the animals go east: they make for the
wadi Greygi and go as far as Ba-hay. Departures take place towards
the end of July, before the heaviest rains, or else at the end of September, depending on the year. The girls leave with the herdsmen
in order to gather 'kreb'. The entire area is exceedingly fertile and
the herdsmen would rather run the risk of losing a few animals in
the clayey areas than take their flocks and herds onto the 'goz' where
the grass /miné/ is far less nourishing, especially for the cows.
Occasionally, the few herds of camels leave for more than six months,
no doubt to join the circuits in the Dar-Fur we have described earlier.

During the dry season, /áĩgì/, in other words from February to
June, the flocks and herds draw closer to the permanent wells.
Those from Oru-ba now enter the region of Uru, a mountain east of
Oru-ba, approximately 40 km from the well where they are watered,

with the cows drinking every three days, the camels once a week. The journey to the wells takes one night and one day, as the animals travel quite quickly, for they are attracted by the water on their way there and by the grass on their return. They drink on the evening of their arrival and immediately afterwards begin their return. The camels, and particularly the she-camels, are watered during the night to prevent their being seen by people who have the evil eye.

The flocks and herds of the Zaghawa of the Chad (even in the case of the Kobé district where the animals seem more numerous than elsewhere) do not travel as far away from their owners' villages and the permanent wells as those in the Sudan. For in the Chad they do not have those vast grazing areas open to pastoral movements. This is the primary obstacle to the enlargement of the flocks and herds and prevents an increase in the number of camels.

Nevertheless long journeys with flocks and herds far from the village are not infrequent, as the following extraordinary story illustrates: Abdullay and Ismaël, two young herdsmen, went off towards the saltings ('sabka') and "the grazing grounds of the oryx and the aryal"[68] with a flock of sheep and herd of goats of about 150 head in their care, together with a group of fifteen young boys entrusted to an older one well acquainted with the country. The two young boys remained absent during five successive years, living on milk and roasted meat. Nearly every week they used to kill a ram for food, while milk and wild water-melons /oru/, plentiful in the dry season, quenched their thirst. To amuse themselves, they played musical instruments, danced and told stories. This untrammelled life turned them into real "brigands" who had no scruple in robbing and killing anyone that happened to cross their "territory". Five years later, on their return home, their flock and herd had trebled and they themselves were fine, sturdy and healthy young men. During the five preceding years their parents had heard from them only once.

HERDSMEN AND THEIR WAY OF LIFE

One should not assume that the Zaghawa herdsmen are in the habit of getting involved in such epic adventures. Their many duties are somewhat more monotonous, though watching over, milking and watering the flocks and herds, differ for animals staying in the proximity of dwellings and for those on the move.

Sedentary flocks and herds

Cattle — The herds of cattle remain close to the dwellings and are usually left to graze without supervision, with the exception of the rainy season when they might stray into the fields and damage the crops. During the other seasons the cattle usually go in the morning to the pastures on their own, after being milked and watered. They return in the evening, at dusk, and are shut up in a circular enclosure formed by heaps of thorn branches (Z. /guli/; Ar. 'zeriba') over a metre in height. Within this enclosure is situated a smaller one in which the calves are kept. Once the cows have been shut in the larger enclosure the women bring the calves to suckle in the following way. For the first two months they pronounce the cow's name while they drag its calf up to it; after this the calf will come on its own as soon as it has heard its mother's name called out several times. The cow's hind legs are tied together quite high up. The calf is first allowed to suckle for five to ten minutes until milk comes, which usually requires some time; then, when the woman sees the milk coming, she begins to draw for herself while the calf continues to suckle.[69] Each animal yields between a litre and a litre and a half of milk on an average. During one year, and sometimes longer, the calf suckles its mother.

In the absence of its calf the cow would not yield its milk but withhold it. Accordingly, when a cow loses its calf, either milking must cease, or else another cow's calf is brought to deceive the reluctant one, sometimes with no results. Occasionally a little natron and salt will succeed in appeasing the unfortunate animal which will then consent to give its milk, but this is rather unusual.[70]

Apart from this system of separate enclosures, the Zaghawa know other methods for preventing kids, lambs and calves from suckling their mothers; for instance they may fit the young calf with a muzzle or else, when an animal is pregnant, smear its udder with cowdung or goat droppings, or tie a small piece of wood onto each of its teats.

Before the cows are released in the morning, the calves are carefully separated and shut up in the inner enclosure, for once a cow has gone to the pasture with its calf, they will never return. It is in the regions overrun by wild animals, which may attack the cattle, that the need arises for sheltering the cows every night in the enclosures. During the rainy season no night grazing is allowed, for fear of the cattle damaging the crops; this is the reason why in all parts they are kept within the enclosures every night till harvest is over.

But when possible the cows are released immediately after the evening milking, to spend the night on the pastures. In that case they do not return in the morning to the enclosures and consequently there will be no milking or watering at this time of day; the cows return only in the evening to be watered and milked. To sum up, whatever the case, the cattle return to the enclosure only once a day, in the evening.

Watering takes place very early in the morning, at daybreak, or else in the evening. When water is plentiful, the cattle drink once a day, but, if water is scarce, every other day or every third day; they may be watered together with the sheep and goats. They know their way to the well and usually go there in a herd on their own, with one of them out in front, apparently leading. The woman or girl in charge of the watering may follow a few metres behind on the path leading to the well, carrying her dipper on her shoulder. But more often than not she is already on the spot, filling the waterskins for the family supply, as she awaits the herd. The wells, dipper and troughs have been described above (see pages 37-39).

We must now emphasize the conditions prevailing in the Guruf district of Chad. In this extreme western corner of Zaghawa country conditions are very special, in that large areas are cultivated with millet. The animals are never allowed to graze without herdsmen. The camels are few but the people of Guruf own large herds of cattle and flocks of sheep. Besides the Guruf abounds in wild animals such as lions, hyaenas and jackals, against which the flocks and herds must be protected.

According to our informants here, in each family the duties are distributed as follows: in the morning, at sunrise, the father opens the two enclosures in which the cows and sheep have separately spent the night. One of the children, either a girl or a boy, having taken some food, goes off for an hour or two with a mixed flock and herd a short distance away from the village. At about 8 o'clock a.m. the child brings back the animals. The mother, sitting on her haunches with a calabash held tightly between her knees, milks the cows with both hands. In the meantime the father has gone off to the well to draw water for family use. When the milking is over, the mother feeds her husband and children; and one of the children, dressed in a short shirt if his father is rich, otherwise just a sheepskin, and carrying a stick, goes off to the well to water the cows, then takes them to the pasture. He may go as far as 3 or 4 km only to return at nightfall. The mother then puts the calves kept in the enclosure to suckle while she milks the other cows; when this is finished she leaves them all shut up for the night, to go and feed her husband and children. After their evening meal, the father goes to sleep, with his dog, his spear and sword, in a circular, roofless dwelling beside the animals' enclosure; the mother sleeps with her children in the hut containing the granaries. Though it rains rarely at night in the rainy season, the father takes shelter under a roof during this period.

Sheep and goats — Unlike cattle, the sheep and goats are rarely left by themselves. The herdsmen may be boys or girls from 6 to 15 years, sometimes brother and sister together. However, nowadays one may see men guarding these animals.

At Um Buru, Kornoy or Tiné, sheep and goats may be brought together in one herd but at Anka, Musbat and Forawiya they are always kept separate: the goats go with the cattle and the sheep with the camels. Sheep and goats are shut up for the night in an enclosure of thorns, not as high as that of the cattle but including, like the latter, an additional small enclosure to separate the lambs and kids. The animals are never released at night with the obvious intention of keeping an eye on them and also for fear of jackals and hyenas who are more prone to attack the sheep and goats.

To milk the goats and ewes, the woman or girl catches the lambs and kids one by one, grabbing its leg or ear, and taking it to its mother. Once it has had its fill, or even while it is suckling, the girl draws milk for the family in a wooden dish or a halved calabash. The woman or child, crouching down beside the animal, raises its right leg which she lays on her thigh or knee and draws the milk. When it does not come easily, she gently punches the animal's udder, in the same way that the kid or lamb would have butted it /dùk ili/ "it struck the udder [of ewes and she-goats]" — this is never done when cows are milked.

These animals are watered at the same times as the cattle (in the morning or in the evening); they are watered around midday when it proves necessary to spread out the arrivals of the flocks and herds at the wells.

Transhumant flocks and herds

The flocks of sheep and herds of camels are accompanied to the 'jizu' by the eldest, unmarried boys and girls, sometimes also by young married men who leave their wives behind in the village. Whether the pasture is near at hand or several days away on foot, the young herdsmen leave alone without dogs (except for the Zaghawa of the Chad where each herdsman has his dog); they are poorly dressed in a short cotton robe or sheepskin and carry a crooked throwing-stick or a staff.

On their arrival in the 'jizu' pastures, they disperse, each one going where he pleases with his flock or herd, keeping at a distance from the rest, alone with his animals. As a young herdsman put it:

"it's better to stay alone than to mix your animals with other people's". From time to time a few of them come together to chat and amuse themselves. Herdsmen build small huts for shelter or simply set up a blanket as a kind of tent; for some time now they have been using thick tarpaulin that comes from Egypt.

Neither men nor animals drink water, apart from the short period when there still remain a few pools. The basic diet of the herdsmen consists of fresh camel's milk, drunk undiluted, and of curdled milk /ó kèrr/.[71] "When you have drunk camel's milk", say the Zaghawa, "your thirst is completely quenched;[72] after cow's, ewe's or goat's milk, you still feel thirsty". They also eat flour balls kneaded with milk and cooked on embers, accompanied by sweet tea. To make their tea they carry waterskins filled with water; when these are empty, two or three people go to the nearest well to refill them. Whereas in the village only women milk the cows, goats and ewes, since the men "do not wish to soil their robe",[73] both men and women milk the camels; but as a rule this task is usually performed by men. In the course of transhumant movements, both boys and girls, circumcised or not, carry out the milking twice a day (morning and late evening). Sometimes they milk the animals a third time to allay their hunger and thirst.

They while away the rest of the time making camel hair into rope, saddlegirths, narrow carpets, caps and belts.[74] Camel hair is cut at any time of year, though usually before the dry and hot season. The herdsmen can also make ocarinas and small clay flutes (both instruments are called /mori/); nowadays they are usually made of metal. They do not play for themselves but for the animals. "Sheep like music", said one of the herdsmen; music may also prevent the flock from straying.

Zaghawa herdsmen know how to treat certain animal ailments by cauterization with a red-hot iron, bleeding, adding natron to drinking-water, smearing tar on open wounds, etc. They can also castrate camels, bulls and rams.

The herdsman's position

There are two kinds of herdsmen. The first category consists of young boys or young men who perform this task on behalf of their own families, without however acquiring a special position. This is especially the case among the Zaghawa of the Chad. Another category comprises the paid workers hired by an owner for an agreed wage, which is frequently the case in the Sudan, where both kinds of herdsmen are to be found however.

When the herdsman is an unpaid member of the family, he is

usually a son, son-in-law, brother or a more distant relative. The Kobé prefer to leave their flock or herd in the care of a Bideyat herdsman, more often than not a distant relative. The flock and herd in his care may be the property of one man or else may comprise animals belonging to various owners, such as a father and his married sons, or several brothers or even sometimes three or four different heads of family who live in the same village, and so on. Any grouping is possible. The herdsman does not receive any fixed remuneration. He is fed, in other words he may drink all the milk he likes; he is also clothed (robe and sandals) and if he proves himself a good shepherd, he will be presented with an animal from time to time. The owner of the flocks and herds usually entrusts him with one of his young children to accompany him.

The hired herdsman /karda/ is not related to the owner of the animals in his care. His wages are usually in kind, not cash. Only the owners of large herds may engage a paid herdsman, for in such a case it is not customary to combine one's animals with those of a few relatives in order to entrust them to the same herdsman.

In the Sudan a camel herd of 60 to 100 animals is regarded as the maximal unit (Ar. 'morah')[75] which one man should supervise alone. For one year of service the herdsman of such a herd receives a young male or female camel, worth £S.12 on an average. A herdsman who looks after more than 100 camels, say some informants, will receive two camels as wages; but, according to others, when an owner has over 100 camels to be looked after, a fairly rare occurrence, he will split his herd and hire a second herdsman. The hired herder receives as his wages eight sheep a year, two of them in advance. The animals received remain in the owner's flock, and if one of them dies, it is not replaced. The herdsman is also given clothing consisting of a robe, a 'tōb' (a large rectangular piece of cotton worn draped like a toga), a pair of sandals and in addition flour, sugar, tea, or else a little money with which to buy some. In the bush all the milk left after suckling by the young camels goes to him. When the herds are near the villages, the owner takes some milk for his requirements of fresh and curdled milk, and of butter; in this case the herdsman's share is reduced to a small amount.

The cattle too may be entrusted to a herdsman. One of our informants at Dor, who owned about 50 cows, had them looked after by a Zaghawa herdsman who kept them together with his own. This herdsman is not only clothed and fed, but disposes freely of the milk and butter. The owner gives him neither animals nor money, but by way of compensation he pays the taxes for all the animals in the herd, including those of the herdsman; and if the latter is in need of money, he is permitted to take an ox from the herd and sell it.

Whatever the herdsman's status (hired or not), the increase in the herd goes entirely to the owner. A sick or hurt animal must have its throat ritually cut before it dies; if not, its flesh is unclean and cannot be eaten. When the owner is close by, the herdsman takes the skin and flesh to him. However when the owner is far away, the herdsman collects witnesses and in their presence cuts out on the carcass the piece of skin where the clan mark /erfe/ has been branded; this must be kept and delivered to the owner. As for the meat, it is his to dispose of.

Of course, being Muslims they do not eat the flesh of an animal which has died naturally; they cut its throat ritually when it is still alive, that is to say they cut its jugular vein as its head is turned to the East. An animal that had died before this could be done would be treated as carrion and could not be eaten. In the case of an animal attacked by a hyena, the herdsman removes the bitten part of the flesh; the remnant is clean food.

Once again we have been able to observe that the Zaghawa of the Chad and those of the Sudan do not resort to the same solutions. Among the Zaghawa of the Chad there is no notion of salaried labour.[76] Family manpower is called upon, a good herdsman being rewarded with the gift of an animal. This is practised as well by the Zaghawa of the Sudan, who have the additional possibility of calling upon somebody from outside the family; such a herdsman receives wages — rarely in cash, it seems, since it usually takes the form of an animal and various goods or victuals. Ultimately the difference lies not in the nature of the herdsman's remuneration, but in the existence of a formal agreement setting out the obligations contracted by the herd's owner. In the former case the herdsman is subject to the good will of the herd's owner, but this is qualified by the fact that an understanding can always be reached between relatives; in the second case the two unrelated parties are linked by a contract.

ANIMAL PRODUCE

Milk — The Zaghawa consume the milk of camels, cows, ewes and goats in various forms: fresh or sour, hot or cold, pure or mixed with water or tea, and also as curdled milk; they prepare milk porridge and know how to make powdered milk. The Zaghawa make butter, but not cheese.

To make curdled milk, /ó kèrr/, fresh milk is first placed over a fire, and allowed to boil; it is then poured into a calabash or earthenware vessel which is hung in a net from the roof of the dwelling, not far from the hearth. It is left there for two or three days before being drunk. Alternatively milk may simply be left in the sun. After a time the thick part floating on the surface of the liquid is removed to be drunk without further preparation; or else it is dried in the sun to obtain a form of powder, /ó savi?/, used by travellers.

Whey, /ó tògòmó/, is drunk by humans and animals. Blacksmiths often come to claim this food and it would be shameful to refuse them. Porridge made by throwing fresh millet flour[77] into boiling milk is called /ó kiláû/. It is mixed with a stirrer (Z. /dìdì/; Ar. 'warwar'). The Zaghawa also drink the beestings of a mother which has just given birth.[78] They are very fond of this fatty and nourishing food.

Donkey's milk is occasionally drunk as a remedy against fits of coughing. For instance a child suffering from whooping cough, /jornok/, is given fresh donkey's milk to drink but great care has to be taken not to tell him the nature of the remedy since he would refuse to drink it.

The Zaghawa use butter in cooking. They also make use of it as a remedy, a cosmetic and a preservative for leather. When butter is made, the various milks are never mixed, as for instance camel's milk with cow's milk or ewe's milk; this is not due to a prohibition but because, according to the informants, "some people dislike certain kinds of butter". The churn is a calabash hung from a tripod and the woman agitates it by repeatedly pushing it with her hand, as she sits on the ground between two of the tripod's legs.

Butter, /boru/, takes two forms: solid /boru deni/ or liquid /boru hamu/. Solid butter is fresh butter; it comes as balls floating in a calabash full of whey; it is administered to babies in the form of small balls; in the latter case it is regarded more as a medicine than as a food. Women use it to smear their plaits.

To obtain clarified butter, /boru hamu/, fresh butter, /boru deni/, is placed over a fire. When it starts boiling, a ball of flour kneaded with water is dropped in; then froth mounts to the liquid's surface. When it disappears, it is time to remove the pot from the fire; the butter is now melted and will remain liquid. The ball of cooked flour, /boru soru/, stays at the bottom of the pot; women and girls share it but boys have no right to it. If one of them did eat some /boru soru/, the others would insult him and a fight would ensue.

Clarified butter is kept in tightly made basketwork vessels or else calabashes for which a basketwork neck and lid have been made, or again in skins made from the skin of a camel's neck, /taga-sao/, or lastly, for some time now, in bottles or tin cans. It is particularly used in cooking, but also in ointments. Women smear it on their bodies, it is put on children's newly shaven heads; it is also used in dressing wounds. It is rubbed on the coats of the horses to rid them

of mosquitoes. It is drunk as a medicine: children who cough or are
unwell are given two or three glasses of it, followed by hot tea. It
may also be given to adults for the same reasons.

Hides and meat — If a two or three year old billy goat or ram is
readily killed, it is rare on the other hand to slaughter cattle or
camels, except for the periods of sacrifices and important cere-
monies. Only men may kill an animal. Women are not permitted to
do so, whether it be a chicken or some other animal; nor may uncircumcised boys do so. In a village where all the menfolk had to go
away for a long period, a boy had to be circumcised, we were told,
in order that he might slaughter animals after the men had gone. As
in all Islam, the act of slaughtering has a ritual character and, as
has been seen, is performed with the animal's head turned to the
east and the blood is left to run onto the sand.

To skin a sheep, ewe, she-goat or billy goat, one blows air under
its skin, applying one's mouth through incisions made around the
legs. When the animal is inflated like a balloon, it is hung from a
high branch and skinned by peeling off its skin inside out like a glove.
It is then emptied and cut up. Of the inside, only the contents of the
rumen and that of the intestines are discarded together with the
large intestine. The remainder is eaten: the small intestines are
immediately grilled on tiny sticks over embers; the liver, lungs and
stomach are quickly scalded, then cut into small pieces and eaten
almost raw with some gall and red pepper as a flavouring (this dish
is called 'marara' — a word meaning gall in Arabic).[79] The liver is
considered the daintiest bit. It is accordingly shared between all
those present in order to prevent ill-feeling on the part of anyone
who does not receive any. Women and children do not eat 'marara':
it is food for men. The kidneys may be cooked separately or served
in small pieces among the 'marara'. The heart too is eaten.[80] The
ribs, fillets and legs are generally grilled; the rest of the meat is
stewed and used to make sauces.

The procedure is the same when cattle are killed: slaughter,
skinning and finally grilling or boiling of the meat; the interior organs
are not eaten. The cows are only killed when they are very old; the skin
is kept and the meat given to the blacksmiths who dry it and sell it
on the market. Nowadays it happens that blacksmiths refuse this
kind of meat.

In the case of camels, a different procedure is observed. Before
being skinned, the animal is cut into pieces. First of all the neck,
whose skin will be used to make a large receptacle for butter, /taga-
sao/, is removed, then the hump. The fat inside it, /boru ergi/,
will be used to smear women's plaits, children's heads and for

various bodily applications; the skin from the hump will make a soft bucket for removing sand from the wells. Next the four legs are cut off. Each part is skinned: millet sacks, /kara/, will be made from the skin of the trunk, shoes from that of the legs. The meat is eaten grilled or boiled.

No one would think of eating donkey or horse meat. These animals are used solely for transport.

Dippers, skins for grain or water, clothing and footwear are made from the hides of various animals; with the hair — mainly camel and goat hair — are made rope, belts, carpets, etc.

Excrements — Cow urine, used in other countries for washing milk receptacles, is only used among the Zaghawa for removing the hair from skins intended to be tanned. The dung is used as a plaster for wounds; added to clay, it acts as a binding agent in the making of earthenware vessels, granaries and house walls. Dung and dried droppings are used for making fires.

CONCLUSION

The empirical yet rational use of water and pastoral resources, revealed through the study of the seasonal movements of transhumance, allows the Zaghawa to maintain and augment sizeable flocks and herds, whose composition varies according to the regions.

The animal resources provide men with a good proportion of their food, clothing and possessions; however, apart from sacrifices, owners avoid killing or selling cattle and camels. These constitute an asset which they develop but do not squander and which they acquire or increase whenever they can obtain money by taking a paid job outside the country. The following words from a song accompanying a dance put the matter as follows: /man báná táîró/ "without any animals, you won't find [a girl to marry]". One must therefore have a herd or flock in order to take a wife or marry off one's sons. One sultan intent on laying down law tried unsuccessfully to reduce bridewealth to ten cows. It is still frequent to find bridewealth as high as forty animals and not unusual to hear of sixty to seventy head of cattle, which remains the goal people try at any rate to attain.

It has often been asked whether the Zaghawi is half nomadic or half sedentary. We would reply that just as in European husbandry he is aware of "the need for transhumance with his flocks and herds, in other words the need to move around according to the seasons in order to find the water and grass necessary for his animals".[81]
Should an end be put to this procedure (incorrectly described as

nomadic), as certain politicians advocate? We regard this form of husbandry as a response to certain ecological factors (size and irregularity of rainfalls, soils, vegetation, etc.). Any attempt to do away with it would be retrograde. For in actual fact all the regions that have less than 200 mm of rainfall annually can only provide temporary pastures where the flocks and herds may spend four to five months each year after the rains. To cease the practice of transhumance would be tantamount to surrendering this vast area to the desert, for it has no other possible use.

During our last field-trip (October-December 1970) we observed that after two years of inadequate rains followed by a year of almost continual drought (1969) the traditional circuits had to be abandoned. A vast number of departures for the south, beyond Zalingei, had taken place, leaving only a few people in each village. Men and animals began the emigration from January onwards; but though people were able to survive and feed themselves by buying cereals or else working as day labourers in order to obtain some, the flocks and herds for their part suffered considerable losses due to insufficient nourishment and, in the case of those that had gone too far south, due to the bites of the tse-tse fly.

In 1970 there were abundant rains. From July onwards the majority of the Zaghawa regained their villages, obviously without having been able to reconstitute their flocks and herds. However, a few, having been warmly received by the southern peoples, had not yet returned home by December 1970.

Original French article, 1971: "Système pastoral et obligation de transhumer chez les Zaghawa (Soudan-Tchad)" Etudes Rurales, 42:120-171.

Chapter 4

TRADITION AND DEVELOPMENT

Indeed one or several successive years of drought, such as in 1969-70, brings famine followed by an exodus from the area and the death of the animals; and it raises an agonising problem for the people: should they remain in so harsh a country or try to settle elsewhere? We experienced one of these periods and were invited by some Zaghawa of Dar-Fur to participate in their discussion of the matter.

As seen earlier, Zaghawa territory consists of a grass and thorn-covered steppe with a mean rainfall of 300 mm in the south and 100 mm in the north, whose showers are centred around August during a brief but irregular rainy season. The Zaghawa practise on a permanent basis an extremely diversified form of transhumant stockbreeding that involves cattle, camels, sheep, goats, horses and donkeys. Now and then they undertake too the cultivation of millet. They gather wild seeds and fruit, and hunt. Lastly they trade, often by barter, in order to obtain those products they are unable to procure in their own territory. They normally manage to achieve a precarious balance that allows them to survive and bridge the harsh periods of scarcity.

After several stays among the Zaghawa, we wonder whether it might not be possible to take some practical measures to ease the way of life and raise the living standards of those breeders living in the arid zones, increasing in the process their contribution to the general prosperity of the country to which they belong. The simple solutions we are putting forward here, which are applicable to other states in the Sudanese zone, are based on our knowledge of the Zaghawa community. We submitted them to some Zaghawa elders, and observed their reactions. As a result we feel we can reasonably affirm that a concrete improvement may be realised with the consent of those concerned and without any need for compulsion. Instead of a sudden and dramatic upheaval involving sophisticated techniques from outside, it would take the form of a long term project, calling upon local practical and scientific knowledge of the environment and utilising the country's own manpower resources. In some cases the project's results would be felt almost immediately, in certain other cases within the next five or six years. Moreover they would provide a basis for further developments.

Our proposals deal with the improvement of pastures, the possibilities for a limited form of new agriculture and the introduction of arboriculture on a small scale, the improvement of horticulture, a more efficient utilisation of water, the development of crafts and home industry, and finally the training of the people.

1. IMPROVING THE PASTURES

As a starting point, we use the botanical knowledge of the breeders. The Zaghawa are well aware that in certain pastures ('jizu') some of the plants are eaten by camels and others by sheep. As a consequence they group these animals and send them off to graze together. They know about the various types of pasture: coarse or tender, more or less salty, tonic, constipating or purgative, irritating, poisonous, more or less nutritious, and so on. They know how to use the browsing pastures provided by certain trees: herdsmen are used to knocking down leafy branches and pods with a wooden tool,/kĩ/, in order to feed their flocks and herds (see page 35). Improvements may be effected then, using this body of technical knowledge as a point of departure.

In our opinion the best plan would be to find a balance between the grasses grazed by sheep and those grazed by camels. Certain types of fodder should be favoured with regard to their degree of appetency.[82] People would have to check the expansion of the poorer kinds of pastures, mainly Cenchrus biflorus (Ar. 'askanit'). Such grasses, left untouched by the animals, are as a result prone to spread at the expense of the richer pasture, consisting mainly of 'absabe' (Dactyloctenium aegyptiacum) or 'kreb' (Panicum laetum, Echinochloa colona, Eragrostis cilianensis, etc.), wherever both types of pasture are competing with each other:[83] because of the animals preference for 'absabe' or 'kreb', these are grazed before their seeds can ripen and so have more trouble in reseeding while the seeds of the despised grasses fall freely on the ground. In the days when the practice of gathering was more widespread, the areas where 'absabe' or 'kreb' were still growing were protected from animals until the grain had been harvested. This situation was considerably more favourable than the present one. A possible way of re-establishing a balance would be to have goats or other animals graze unripened Cenchrus biflorus, using thornbush enclosures. On the other hand people should go on gathering dry Cenchrus biflorus seeds to make tar.

A systematic effort should be made to destroy quite useless shrubs like Calotropis procera (Ar. 'ušar') which rapidly reproduces itself

over the sandy stretches and turns them into irreclaimable deserts
by eliminating all other vegetation. In their place should be planted
trees that provide browsing pastures, such as Maerua crassifolia
(Ar. 'kurmut'), Balanites aegyptiaca (Ar. 'hejlij'), Acacia senegal
(Ar. 'kitir').[84]

Improving the pastures along these lines has the advantage of
operating within the framework of the present pastoral system, which
is a rational system based upon the transhumance of part of the animals.[85]
The transhumance in question should be maintained and extended if
possible; on the one hand it allows for a diversified raising of herds
and flocks, with camels and sheep on the move for four or five months
in the year while cows and goats remain close to the villages; on the
other hand it offers men the opportunity to exploit resources in areas
which, were it not for their frequent comings and goings, would be
abandoned to the desert without any advantage accruing to the community.

The possibility of storing grass should not be neglected either.
We came upon an instance of it in Kobé Zaghawa country at the village
of Séli-mara, but do not know of other examples elsewhere: this does
not mean it was a unique case. Cut with a scythe straight after the
rainy season, the grass is put to dry and stored on the branches of a
tree. This method could be made more widespread so that the resulting stocks of dry fodder could be consumed during the dry and hot
season, when the animals are gathering around the wells. This technique is more effective than that of allowing standing grass to dry,
as it protects the fodder against animals and fires.[86]

2. THE POSSIBILITIES FOR MODIFYING AGRICULTURE

The Zaghawa know how to grow millet. They prepare the fields and
roughly fence them round with thorn bushes. Then they sow, hoe the
young shoots, drive away the birds and finally harvest the crop. This
series of operations is spread over the period between May and
October. There will be a harvest if there are sufficient rains. Otherwise the ears will not ripen and sometimes less than the original
amount of seed will be gathered. Consequently no sowing takes place
in those years, when the rains are late in coming. This hazardous
aspect of the cultivation of millet, a by no means difficult cereal,
shows us the extent to which growing millet is a gamble.

In addition the Zaghawa gather the wild grains, of which there is
a large variety in their country, and indeed used to gather them in
larger quantities sometime ago.[87] These wild grasses are far less
demanding than millet. It is this situation which has led us to suggest

to the Zaghawa — and to peoples living under the same ecological
conditions — that they should sow these grasses on suitable soils.
Failure is unlikely and eventual loss is reduced to a minimum. These
grasses have the further advantage of being available from the end
of August, whereas at best millet does not ripen before October.
Since 'absabe' and 'kreb' may both equally make excellent polenta, it
is these latter grasses which, to begin with, would be best cultivated,
in as much as each needs a different type of soil.

This attempt to establish the cultivation of cereals that have
already adapted themselves to the local conditions can go hand in
hand with another possibility for improvement through the selection
of their seeds. At this stage, some kind of specialised institute
might perhaps be involved. This, of course, does not imply that the
Zaghawa should give up millet cultivation.

3. INTRODUCING ARBORICULTURE

Men and animals obtain from a large number of trees a supplement
of food, particularly appreciated in periods of shortage, consisting
of fruits or leafy branches and pods, respectively: 'korno' (Ziziphus
spina-christi), 'hejlij' (Balanites aegyptiaca), 'moxet' (Boscia
senegalensis), to give only a few instances; these bear fruits which
are eaten fresh or from which sugar or flour are extracted and whose
kernels are used too.[88] The Acacia seyal ('talha') and the Maerua
crassifolia ('kurmut') constitute excellent browsing pastures. The
Acacia senegal ('kitir') produces gum which, when put on sale,
injects new money into the economic circuit.

These trees are scattered in the bush at the mercy of depredation
by men who do not hesitate to break the boughs when they need them,
sometimes irremediably damaging the tree. Damage by the animals
is also heavy, especially on the young trees which they may eat up
almost entirely and prevent from growing.

A start to arboriculture ought to be encouraged through the setting
up of small plantations of indigenous trees belonging to the same or
associated species so as to promote a greater density of individual
trees. While the young trees are growing, the plantations in question
should be protected by means of fences made from heaps of thorn
bushes ('zeriba') — a method well known to the local people and used
for fields and animal enclosures.

A slightly more risky experiment that might be attempted is to
extend the cultivation of the date-palm whose fruits have a high
nutritive value. The Tunjur, the southern neighbours of the Zaghawa
of the Sudan, have been practising this form of cultivation for a very
long time. Indeed in the regions of Kutum and Fatta Bornu the beds

of the wadis look like palm-lined streets. Numerous plantations are
also to be found in the mountains of the Dar Furnung, by the springs.
A few Zaghawa could undergo a technical training with the Tunjur.
Shoots might be transplanted on an experimental basis to various
carefully selected spots; there could be a first trial in the regions
of Dor and Anka with some chance of success, in our opinion,
provided the young trees are watered during the early years. The
type of palm tree grown by the Tunjur might furthermore be improved
with the help of a specialised institute.

4. IMPROVING HORTICULTURE

In the wadi beds and on their banks women are accustomed to set up
temporary gardens in which, after the rains, they grow tomatoes
(subsequently dried), pimentos, onions and okra (Hibiscus esculentus). These gardens, which provide the basic ingredients for the
sauce which goes with the daily polenta, do not appear to receive
much care or attention. They could give much better results without
a great deal of effort. In view of their position close to the wells,
channel irrigation on the Tunjur model should be proposed and encouraged. As a matter of fact in their own gardens the latter use
simple but effective techniques of irrigation. Accordingly it should
be possible to maintain permanent gardens till almost the end of the
dry season.

It might also be possible to introduce new vegetables that could
become a natural part of the daily diet, like aubergines (egg plants),
courgettes or pumpkins, beans, chick-peas, and haricot beans, or
again fruit trees like the orange, lemon, mango and guava trees,
just as we saw in the governmental gardens and those of a few local
chiefs.[89] An attempt should be made to acclimatise varieties such as
at present flourish in Africa.[90]

5. THE WATER PROBLEM

The techniques for drawing water and storing water may be improved.

A. Drawing water

To this very day water continues to be drawn by hand, using a dipper
hanging at the end of a rope. As a rule wells reach a depth of between
15 to 20 m, although certain permanent wells may attain as much
as 60 m. This means an important human expense of time and effort.
Thus one can easily understand how people, who are principally herds-

men, are reluctant to go on drawing water to irrigate their gardens, once they have watered their animals and seen to the family needs.

The colonial administrations had begun to construct cement wells that were equipped with diesel-engined pumps or metal norias ('sakia') operated by a handle, as seen at Hiri-ba in 1956. In this particular village the handle found its way to the bottom of the well within a very short time; after turning the wheel with their own hands for a few days, the users of the well returned to their leather dippers and ropes.[91]

It should be remarked that in general the local people using the wells fitted with diesel-engined pumps are not in a position to buy the fuel at their own expense and that there is no skilled mechanic on the spot ready to repair the equipment in case of breakdown. Such incidents could prove disastrous in the dry season, when large numbers of flocks and herds are gathered together.

Perhaps the most suitable course would be to introduce the population of these regions, consisting after all of herdsmen, to the kind of animal-drawn well which may be seen all over the central Sahara, in the Fezzan, the Tassili, the Air and the Adrar to give only a few places. R. Capot-Rey gives a detailed description of it which is worth quoting here:

"It's an ingenious method of drawing water which replaces man's vertical pull with the horizontal pull of an animal, a mule or camel.[92] The system consists of a 'delu' (dipper) open at one end, like a bucket, and ending at the other with a pipe. The 'delu' comes up with the pipe, held up by a rope which runs over a pulley. When the pipe reaches the level of the roller fixed below the pulley, it falls back

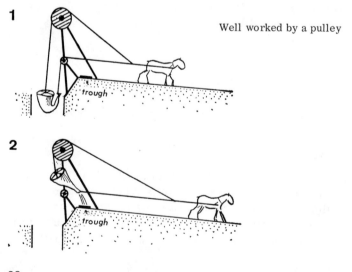

Well worked by a pulley

and the water runs into a trough. The animal's effort is reduced by building the towing path with a slight slope so that when the animal is pulling up the 'delu', he is descending and when the 'delu' is going down, he is climbing. The 'delu' holds between 20 and 30 litres of water and goes down twice a minute on an average."[93]

In the course of the discussion on the problem of wells, a prominent Zaghawa told us that he had heard of the existence of this kind of well among the neighbouring Berti.

A further possibility is offered by another method, the shadoof, found in the Fezzan, the Ahaggar, the Adrar as well as the Tibesti, the Sudan and Egypt.[94] Once again we quote from R. Capot-Rey:

"The shadoof consists basically of a lever turning about a horizontal crossbar fixed to two upright supports of wood, brick or cement. The lever is made from the trunk of a palm or occasionally from two tamarisk poles that can slide alongside each other, following variations in the level of the water. A weight is fixed at the end of the shorter arm and the vessel which will serve as dipper is hung from the end of the other arm. The well is manually operated by pulling the rope at the end of which hangs the dipper which the weight can raise up on its own. The effort involved consists in pulling down the pole and emptying the dipper."[95]

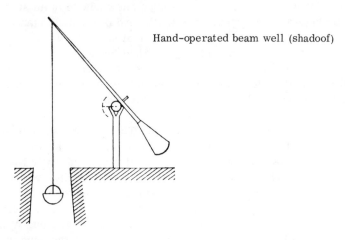

Hand-operated beam well (shadoof)

Both methods offer a certain number of advantages. In the first place construction and maintenance require no cash expenditure. All the material may be found on the spot and maintenance and repairs are within the ability of the users without recourse to outside help.

The animal-drawn well is better suited to great depths. It requires the construction of a sloping bank of earth. The water table must be sufficiently abundant to allow the drawing of a large quantity of water for a number of hours. The shadoof needs longer poles; it draws water from shallow deposits and can manage with a smaller flow of water. However more human effort is involved. "The advantage of using a pulley is that it allows a very long rope to be used whereas the maximum effectiveness of the beam is limited to twice the length of each of its arms."[96]

In practice both kinds of well exist side by side in a large number of cases in the Sahara. A choice then should be made in the light of the local conditions.

With both types water flows into a trough made from cement or wood. In this way water is prevented from becoming sullied by the dipper or the can found in its place, which are customarily placed on the ground, in the mud soiled by the animal excreta. However an attempt must be made to educate people in this matter in order that the trough should be properly situated and used.

B. The storage of water

We know from archaeological remains in the region of the Jebel Uri that the inhabitants of those parts once knew how to build stone dams for storing water. Furthermore the legend of the Judé clan tells us about the construction of a dam, either of stone or wood.[97]

The Sudanese government recently had two small dams built in Zaghawa territory at Ba-sao in Kobé country, not far from the Chad border, and at Metel-Koru, south of Um Buru, in Tuer country. We know a little more about the latter, which was built with local labour and materials, except for the cement, following the plans of an engineer, who was himself a Zaghawi. This enterprise may serve as a model. If this kind of small dam holding water all the year round could be repeated in every place where it was technically feasible (namely a clearly outlined valley with a young profile crossed by a rocky sill), it would be well worth trying. These reservoirs would have to be small and numerous so as to avoid too great a concentration of men and animals during the dry and hot season. With the former there is the risk of brawls, with the latter the danger of overgrazing, not to mention that of epidemics that can affect either group. In our opinion one should strive for an increase in this type of construction rather than aim at more grandiose enterprises. Pumping stations for instance show an unreliable return because of their greater costs, besides involving many risks, such as breakdowns, epidemics and fights.

As a second stage one might consider the possibility of using a part of these water reserves to irrigate the garden cultivations. These waters contain silt and hence are more fertilising than well water.

6. CRAFTSMANSHIP AND SMALL SCALE INDUSTRY

The natural consequence of improving stock-raising techniques in the various ways we have suggested is the enlargement of herds. An increase in numbers would help to make the stockbreeders, who already have sound ideas on the subject, aware of the need to improve the quality of their animals as regards both milk and meat. However this will only come about when a certain threshold in numbers has been attained for, in the kind of society under consideration, in a situation of scarcity the stockbreeder is only concerned with quantity, with quality taking second place.

However there is no question either of aiming at an ever increasing maximum figure of head of cattle, but rather an optimum number which can be roughly defined as follows: the maximum number of animals that the country can feed and water by the end of the dry season in an average year. The optimum number would be empirically defined for each region and found by successive approximations.

It would have to be made clear to the stock-breeders that it is in their own interest to sell their surplus beasts at the beginning of the dry and hot season (February-March) when they are in the peak of condition rather than allow them to die at the end of this season in a rawboned state of thirst and starvation.

A small local organisation might offer to purchase them for slaughter, while taking care not to deprive the flocks and herds of their best animals, for this would run counter to the policy of qualitative improvement which ought to be pursued. The animals bought could be slaughtered on the spot (which implies the setting up of a small slaughterhouse) and the carcases exported by air-freight to the towns. This would avoid long journeys for animals on the hoof to the centres of consumption, with ensuing losses in weight and numbers. The technique of smoking beef or camel meat might also be made more widely known, all the more so since the Zaghawa are already familiar with the method of cutting meat into narrow strips and drying it in the sun. There exists a very simple Ethiopian method for smoking meat (Amharic /qwanṭa/: spiced, dried and smoked beef or mutton flesh, which is as tasty as 'basturma' or the famous "viande des Grisons" from Switzerland). Small factories might be set up at little cost employing local labour in periods when it is relatively unoccupied.[98] Since the majority of the peoples under consideration are Moslem, smoked meat from animals slaughtered by Moslems could be eaten in all the countries that observe the Islamic religion.

The hides might be dried for export or tanned — on the spot or in a place better provided with water — by artisan tanners using local methods, or by industrial tanning.

Blacksmiths ought to be encouraged to take up again a practice only recently discarded, namely the tanning of sheepskins with the wool still on in order to make the sewn skins into rectangular blankets; these are warmer than the imported cotton blankets, last longer and are surely no dearer. Besides local use such blankets could be exported to the neighbouring towns and as far away as Europe (to be used as carpets or bedcovers).

Since the Zaghawa blacksmiths are often also cotton weavers, the latter craft might be brought back into favour and be further improved by the construction of looms based on the same principles as the former ones, but permitting the weaving of cotton strips, 60 or 80 cm in width instead of just simply 10, 20 or 30 cm. Such looms are currently found in Ethiopia.

7. EDUCATION

Whether it is a question of improving those methods and techniques that have been practised for ages, or borrowing certain novel but simple techniques from peoples with a similar type of culture, the users must freely accept these changes and this would be only possible in the majority of cases after appropriate explanations and instruction.

Like many other African societies, the Zaghawa possess a rich source of manpower which should be of benefit to their countries. It would therefore be necessary to set up "rural schools" as soon as possible with the aim of turning herdsmen into technical experts in stock raising, besides training specialists in all the fields mentioned above. These schools should have specially designed syllabuses, textbooks and teaching methods.[99] Since the school year cannot but follow the seasonal movements of the population, the best plan would be to establish an academic year of six months; each three month period spent at school should be followed by a period of the same length spent in the family to avoid too great inconvenience to the latter with the departure of one or more of its members; the school children would not feel cut off from home and their participation in the daily family tasks would not cease.

An experimental school might be opened with the support of the

F.A.O. and a teacher assisted by a technical expert put in charge.
Depending on its success several schools of this type might be contemplated and, after several years, an "Institute of stock-breeding and rural economy in semi-desert conditions" might be established for training engineers. Such schools would naturally be open to all the children wanting to join, whether they were the sons of princes, ordinary people or blacksmiths. The presence of the latter would however raise a few problems. For such is their status that there are very few blacksmiths who would nowadays dare to send their children to school, let alone afford or contemplate it. It is a fact that these children will continue for some time yet to be the only workers prepared to perform tasks which the rest of the population is too prejudiced to undertake.

8. CONCLUSION

Between 1960 and 1970, Zaghawa country suffered several successive years of drought, culminating in the famine of 1969, which resulted in men and animals going south together in search of grass and water. Men and women alike took jobs as paid labourers and were able by this means to obtain cereals. Many animals perished, in particular from tse-tse bites. In Zaghawa country a few animals survived by browsing the branches of those trees that had remained green right through these hard times.

That situation, together with the fear of its recurrence, has incited some Zaghawa leaders to reflect and work out a plan for the definitive transfer of the entire population to the region of Hofrat en-Nahas in southern Dar-Fur (south of the 10^0 N. parallel). The scheme has had a mixed reception, being almost totally accepted by the younger people, while opposed by the middle-aged and the old. A compromise proposal was considered, giving everyone the choice of staying behind, which preserved the existing chieftaincies, or moving south. In any case those leaving were to continue obeying their respective chiefs who would delegate representatives.[100] As 1971 was a better year, the scheme did not come into operation. Only a few persons remained in the south, and their number, it appears, was no greater than that of the regular emigrants who leave Zaghawa country each year in search of supplementary resources.[101]

It nevertheless remains true that a real problem has arisen well worth consideration. One can imagine what would have happened if the Zaghawa community as a whole had gone and settled in Southern Dar-Fur. The result would have been one of the following: either their country would have been turned into a desert or else, more likely, neighbouring nomads like the Kababish or Ziyadiyeh would have added it to the area of their regular circuits, thus making it impossible for the Zaghawa to return to their former villages. These transhumant stock-breeders would have had to become sedentary farmers. How would this transformation have taken place? After all, the region where it was planned that they should settle is already inhabited. How would the local population have reacted to the arrival of such large numbers of outsiders? In time of famine a certain spirit of co-operation may prevail but it is quite another matter to agree to other people settling in your own country. It is fairly obvious that much friction would ensue.

Only if any other solution proves impossible, should one envisage such a sweeping and irreversible measure as the total transfer of a population, or its permanent concentration. A better course would be to improve the present economic model in every way, so as to enable men and animals to live better off in ordinary years and to withstand the years of drought free from anxiety. Should a disaster occur, which always remains a possibility, state aid in the form of grain and eventually fodder would certainly involve less trouble and expenditure — from an economic and social point of view — than transferring the population. From a human point of view, need one stress that a transfer of this nature would, sooner or later, bring about the death of a culture and the impoverishment of a nation?

Original French article, 1975: "Tradition et développement au Soudan oriental: l'exemple zaghawa", in Th. Monod (ed.), Pastoralism in Tropical Africa, London, O.U.P.: 468-486.

Appendix

APPETENCY INDEX

compiled by

PIERRE BOURREIL
MUSTAFA MOHAMMED BAASHER
AHMED MOHAMMED MUSSA

Plants pastured or browsed: 0 = not at all; 1 = very little; 2 = a little; 3 = fairly well; 4 = much; 5 = very much.

Latin name	Sheep	Goats	Cattle	Donkeys	Horses	Camels	Arabic name
Abutilon	0	0	0	0	0	0	niyāda
Acacia albida							haraz
young	0	2-3	0	0	0	5	
fruits	2	4	5	0-1	0	5	
Acacia mellifera							kitir
plant	0	4-5	0	0	0	4-5	
fruits	1	3-4	0	0	0	4-5	
Acacia nilotica							sunut
young	0	4	1	0	0	4-5	
fruits	0	2	3-4	0	0	3-4	
Acacia nubica	0	?	0	0	0	4-5	laɛut
Acacia senegal							hašāb
plant	0	1	0	0	0	4	
fruits	3	3	1	0	0	4	
Acacia seyal							talha
plant	0	3	0	0	0	3-4	
fruits	3	4-5	2	0	0	3-4	
Acacia tortilis							seyāl
plant	0	3-4	0	0	0	5	
fruits	3-4	4-5	4	0	0	5	
Acanthospermum							
hispidum	0	0	0	0	0	0	xorāb hawša
Achyranthes	0	0	0	0	0	0	xašim al ɛama
Albizzia							
plant	0	0	0	0	0	3	abu hirēbat (?)
fruits	0-1	0-1	0-1	0	0	3	arad
Albizzia							
anthelminthica	0	2-3	0	0	0	4-5	kitir
Alysicarpus	1	1	3	2	5	1(?)	abu nigēgēra
Amaranthus							
graecizans	0-1	0-1	0	0-1	0	0-1	tamalēga
Andropogon							
gayanus	2	2	5	3-4	3-4	2-3	abu raxis
Anogeissus							
schimperi	0	0	0	0	0	3-4	sahab
Anthephora	0-1	0-1	5	4-5	4-5	3-4	

Latin name	Sheep	Goats	Cattle	Donkeys	Horses	Camels	Arabic name
Aristida							
adscensionis	1-2	0	2-3	2-3	1-2	4	gau
Aristida funiculata	1-2	0-1	2-3	2-3	1-2	3-4	gau
Aristida mutabilis	1	0	2	0-1	0-1	2	gau
Aristida pallida	0-1	0-1	0-1	0-1	0-1	1-3	simēma
Aristida papposa	1	1	3	3	2	4	nisā
Aristida rhinochloa	1-2	0-1	3-4	3		2-3	4	gau
Aristida stipoides							agēg al bagar
young	0	0	2-3	2	1	3-4	
adult	0	0	0	0	0	1	
Arum sp.	0	0	0	0	0	0	umm barko
Asparagus	0	0	0	0	0	0	umm mušbat
Balanites aegyptiaca							hejlīj
young	0	4	2	0	0	5	
fruits	0	4	2	0	0	5	
Bauhinia rufescens	0-1	4-5	3-4	0	0	5	kulkul
Blepharis linariifolia							bigēl
dry	0	0	0	0	0	3-4	
green	2-3	2-3	3-4	0-1	0	3-4	
Boscia							
senegalensis (bush)	0	0	0	0	0	3-4	muxīd
Brachiaria sp.	2	1	5	4	5	3	abu jigra
Brachiaria	2-3	2-3	4-5	4	4-5	4-5	kurēb
Cadaba glandulosa	?	2	3-4	0	0	4-5	kurmut
Calotropis procera							
(toxic)	0	0	0	0	0	0	ušar
Capparis decidua	0	0	0	0	0	?	tuntub
Capparis tomentosa	0	2-3	3	0	0	4	mardo
Cassia mimosoides							sakarnaba
young	0	0	0-1	0	0	?	
adult	0	0	0	0	0	0	
Cassia occidentalis	0	0	0	0	0	0-1	kawal
Cassia sp.	0	0	0	0	0	0	sanasana
Cassia tora	0	0	0	0	0	0	kawal
Cenchrus biflorus	1	1	3-4	2	3-4	3-4	haskanīt
Cenchrus ciliaris	3-4	3	5	4-5	5	4-5	haskanīt
Cenchrus prieurii	2	1-2	4	3-4	3-4	3-4	haskanīt na ɛm (?)
Chloris gayana	3	2	5	4-5	5	3-4	afan al xadīm
Chrysophora	1-2	1-2	0-1	0	0	3-4	urgasī
Chrysopogon							
aucheri	1	0-1	3-4	2	1	2	
Chrysopogon sp.	2	2	3-4	3	2	3	nāl or marhabēb
Cissus							
quadrangularis	0	0	0	0	0	0	salaɛlaɛ
Citrullus							bitēx
fruits	2-3	3-4	3	3-4	0-1	4-5	
plant	0-1	0-1	0	0-1	0	0-1	
Combretum (bush)	0-1	5	5	0	0	5	šuxiyat
Commelina	2-3	2-3	4-5	1	1-2	2-3	biyēd
Commiphora							
africana	0	2	0	0	0	4	gafal

Latin name	Sheep	Goats	Cattle	Donkeys	Horses	Camels	Arabic name
Corchorus olitorius	0	0	1	?	0	0	muluxiya
Cordia	0	0-1	0-1	0	0	?	indarāb
Cymbopogon nervatus (?)	0	0	0	0	0	0	marhabēb
Cynodon dactylon	3	3	4-5	4-5	4-5	3-4	najila
Cyperus	0-1	0	1	1	0	4-5	umm tuk
Cyperus nitida	0	0	0	0	0	0	saɛada
Cyperus sp.	1	1	0-1	0-1	0	0-1	umm tiwēgyāt
Dactyloctenium aegyptiacum	0-1	0-1	5	2-3	5	0-1	abu sābiɛ
Dalbergia melanoxylon	0	0	0-1	0	0	4-5	bābanus
Dichrostachys sp. (glomerata ?)							kadād
plant	0	1	0	0	0	2-3	
fruits	3	3	4	0	0	2-3	
Digitaria	1	0-1	2	2	2	1	umm ɛag
Eragrostis aspera	2	1	2-3	2-3	2-3	1-2	banu
Eragrostis ciliaris (perennial)	1-2	1-2	3	3	2-3	3	banu
Eragrostis sp.	1	1	3	3	3	2	banu
Eragrostis tremula	0-1	0-1	0-3	0-3	0-3	0-2-3	banu
Euphorbia	0	0	0	0	0	0	umm ɛabaka
Farsetia ramosissima	3	2	0	0	0	?	norām
Ficus	0	0	0	0	0	?	jumēz
Fimbristylis	0	0-1	0-1	0	0	0	umm fisēsyāt
Gossypium	0	1-2	0	0	0	?	gutun
Grewia	0	2-3	2	0	0	2-3	abu wīhīdu, gerēgerām
Grewia ? (bush)	0	3-4	0-1	0	0	2-3	gudēm
Grewia ? (bush)	0	2-3	0	0	0	2	tukā
Grewia tenax ? (bush)	0	2-3	0	0	0	2	gerēdān (?)
Guiera senegalensis	0	0	0-1	0	0	3-4	gubēš
Hibiscus esculentus	0	0	0	0	0	0	bāmya
Indigofera	?	?	0	0	0	3	ašrut
Ipomoea	3	3	3-4	2	0-1	4-5	hantut
Ipomoea repens	0	0	0	0	0	?	arkala
Kalanchoe	0-1	0-1	0-1	0-1	0-1	0-1	
Kyllinga	2	?	1	1	0-1	?	umm tuwēgyāt
Lannea humilis	0	0	0	0	0	?	layūn
Leptadenia pyrotechnica	0	0	0	0	0	0	merx
Leucas	0	0	0	0	0	0	asal al tēr

Latin name	Sheep	Goats	Cattle	Donkeys	Horses	Camels	Arabic name
Maerua crassifolia	0	2-3	5	0	0	5	surēh
Mollugo cyclocarpa	2	?	3-4	?	0	?	garen
Momordica balsamina	0	1	0	0	0	5	ēyer
Monsonia senegalensis	2-3	2-3	4	0-1	0-1	3-4	garen
Ocimum	0	0	0	0	0	0	umm rihān
Oldenlandia	1	0	1	1	0	?	garajub
Oscigonum	0	0-1	0	0	0	0-1	umm hāmdī
Panicum (annual)	0-1	0	1-2	1-2	1-2	1	umm furāw
Panicum turgidum	0-1	1	2	1-2	0-1	3	tumām
Papilionaceae	0	0	0	0	0	0	keldos
Pappophorum	1	1	2-3	1	1	2-3	akariš
Pennisetum	1	1	1	1	1	1	umm dofūfū
Portulaca	0	0	0	0	0	0 ?	darat al baggara
Schoenefeldia gracilis	2	2	4-5	3	3	3	dannab al nāga
Seddera		2				3-4	singid
Sehima ischaemoides	0	4-5	0-1	0	0	5	humēd
Sesamum alatum	0	0	0	0	0	2	simsim
Sesamum indicum	0	0	0	0	0	?	simsim al jimāl
Setaria pallidefusca	2	2	2-3	3	2	1-2	dannab al fallu
Setaria verticillata	1	1	1	1	1	1	umm abaka
Solanum (albidum)	0	0	0	0	0	?	gubēn
Sporobolus festivus	2	2	1-2	1-2	1	?	umm dibēgū
Stereospermum Kunthianum	0	2	0	0-1	3	3-4	xešxāš
Tamarindus indica	0	0	0 ?	0	0	4-5	aradēb
Tephrosia nubica	0-1	2	0-1	0	0	3-4	ašrut
Tephrosia uniflora	1	2-3	0-1	0	0	4-5	ašrut
Tribulus terrestris	3	3	1-2	1	0	2	darēsa
Trichodesma africanum	0	0	0	0	0	?	
Ziziphus mauritiana							nabag
plant	0	5	0	0	0	4-5	karnuya (?)
fruits	0	5	1-2	0	0	4-5	
Ziziphus mucronata	0	0-1	0	0	0	1	nabag al fīl
Ziziphus spina-christi	0	5	2	0	0	5	sidr
Zornia diphylla	3	3	4-5	2	2	2	lusēg or šelīnī

NOTES

1. The ethnological research upon which this book is based was conducted in the course of four field-trips sponsored by the C.N.R.S. (National Centre for Scientific Research): to the Chad in 1956-7 under the auspices of the "Mission du C.N.R.S. aux Confins du Tchad" and to the Sudan in 1965, 1968-9 and 1970 under the auspices of the "R.C.P. 45" (Co-operative Research Programme No. 45).
   Approximately 30,000 Zaghawa live in the Chad (the 1953 census recorded 24,000); there are well over 200,000 of them in the Sudan (230,706 were recorded in the census of 1956 and about 255,000 in 1970). The Bideyat themselves can be very roughly reckoned to be about 15,000 in number.
2. See Joseph Tubiana, "Moyens et méthodes d'une ethnologie historique de l'Afrique Orientale", Cahiers d'Etudes Africaines II, 1(5): 5-11.
3. According to meteorological observations made at Kutum, Um Buru, Hiri-ba and Fada. See Chapter 3, pages 32-33.
4. See Chapter 2, page 26.
5. See Chapter 3, pages 34-36.
6. See Marie-José Tubiana, "Le marché de Hili-ba", passim.
7. Here are a few examples of prices observed among the Zaghawa of the Sudan in November 1970: camels, between £S.15 and 30 or 35 (according to age); horses, £S.10 to 18; cows, £S. 5 to 12 (according to age); sheep, £S.2 to 5; goats, £S.0.50 to 0.75; chickens, £S.0.10 to 0.15.
   One 'šuwal' of millet (100 to 110 kg), worth £S.1.50 in November 1970, had reached £S.4 during the dry season, and £S.7 after the poor harvests of 1969; one 'rotl' (445 g) of sugar, £S. 0.10; of natron, £S.0.03; of salt, £S.0.03; of peppers, £S.0.10 and of onions, £S.0.03.
8. As early as the 19th century, El-Tounsy noticed that, among the Fur, the peasants' wives, who helped their husbands in agricultural work, would nevertheless make provisions of fruits and several wild seeds: 'hejlij', 'nabak', 'korno', rice, 'difra', 'kreb', 'andarab', 'ardeb', 'mokhet', (Voyage au Ouadây, pp 358-359).
9. We are indebted to H. Gillet for the identification of some of the plants in our herbarium. Thanks to his record of the vernacular names of plants in Bideyat, Daza and Chadian Arabic, his "Catalogue raisonné et commenté des plantes de l'Ennedi" has allowed us to identify others. We are extremely grateful to him for having kindly read the French draft of this chapter and suggested some amendments.
10. See Marie-José Tubiana, Survivances préislamiques, pp 170-171.

11. Zakaria Fodul (a Zaghawi from Chad) gave us the following information: in Zaghawa country people harvest more wild /aìrì/ and /jìgírì/ than cultivated millet, and consider that these two wild cereals make a better porridge. This is why they call them both /gú/, whose proper meaning is "porridge" and, by extension, "food".
12. Possibly because they have to be consumed fresh.
13. P. Bourreil identified as Eragrostis aspera a grass he collected in Dar-Fur with the local name 'banu' (Dossiers de la R.C.P. 45, No. 4, p. 24).
14. According to Isa Hasan Khayar, who comes from Abbéché, in the Wadday, 'belile' or 'am-belile' denotes millet which the shepherds or young students of Koranic schools put in a wickerwork basket or an earthenware vessel and soften by adding water before eating it raw. It is also the food of travellers and thieves.
15. Other seeds are used to make tar, like the seeds of /sundu/, an unidentified plant producing grain the size of millet (perhaps Schouwia purpurea Forssk.; Gillet, No. 27?) or the seeds of the castor oil plant (Z. /eigo-biʀi/ "dark brown tar"; Ar. 'kilwe' or 'xirwe'; Ricinus communis L. var. minor; Gillet, No. 134).
16. Marie-José Tubiana has already drawn attention (Survivances préislamiques, p. 101) to the fact that certain Zaghawa clans, among them the /turoŋda/, were in the habit of tearing open anthills to remove the cereals hoarded by the ants in order to eat them, whereas in other clans — the majority of whom are Kobé clans — "ants' seeds" were subject to a prohibition; also that Azza women (a caste of blacksmiths among the Teda and Daza) often seize the hoard made by the ants and were made fun of by the non-Azza. This custom has also been observed among the Maba (Le Rouvreur, p. 171) and among the Tuareg, where "this harvest is regarded as a somewhat degrading act by men of good birth" (E. Bernus, p. 41).
17. J. Chapelle, Nomades noirs du Sahara, p. 382.
18. For this same plant H. Gillet records the Arabic name of 'am zerar' (?) and the Bideyat name of 'nali' which corresponds closely to our /ňáʀì/. He adds that in September this shrub produces "small orange-coloured fruit with large stones and a thin, scented flesh, which are eaten like sweets"; however the fruit, which the Zaghawa call /ňáʀì/, does not have large stones but contains two seeds.
19. H. Gillet (No. 205) gives 'gunda'(?) as a Bideyat name; on the other hand he has also collected in Bideyat the term 'keida' for the Ziziphus spina-christi (No. 206).
20. The Tuareg also eat this fruit and use its flour in a similar way (E. Bernus, p. 42).
21. H. Gillet gives the Bideyat name 'norda'; 'surei', unknown to the Zaghawa, is the Arabic name in Ennedi.
22. Marie-José Tubiana, "Le marché de Hili-ba", Table 1, note 3.
23. The production of gum in Zaghawa country (Chad), up to the 10th of June 1957, amounted to 4,700 tons (J. Barat, "Note sur la production de la gomme dans le district de Biltine", Mémoire de fin de stage, ENFOM, 1957-58, (No. 3):18). We lack details on the collection of gum by the Zaghawa of the Sudan.
24. J. Chapelle, Nomades noirs du Sahara, pp. 191-95.
25. Ch. Le Coeur, Dictionnaire ethnographique téda, see under 'forkoso', "travail".

26. Op. cit., p. 71.
27. Op. cit., see under 'abor' and 'yi'.
28. A. Le Rouvreur, Sahariens et Sahéliens du Tchad, pp. 102, 114, 130, 131, 141, 148, 160, 171, 191 and Index. See n. 31 for the Tunjur of the Sudan.
29. E. Bernus, pp. 33 and 48.
30. We have not been concerned here with hunting, important as it is for the Zaghawa of the Chad. Fishing is, of course, non-existent.
31. In the Sudan, we saw various wild berries offered for sale in the markets during Ramadan. Among the Tunjur of the Dar Furnung we observed large stores of wild cereals, mainly 'absabe'.
32. Further north in the Ennedi, where the Bideyat live, figures are lower. The figures obtained at Fada were: 101.5 mm in 1963 (of which 66.1 mm fell in eight days), 135 mm in 1964.
33. See C. McRamsay, The forest ecology of central Sudan; J. G. Lebon, Land use in Sudan; P. Quezel, Flore et végétation des plateaux du Darfur Nord-occidental et du Jebel Gourgeil, pp. 10-11.
34. 'goz' means a sandy soil, or occasionally a stabilised sand dune. M. Baumer gives among other definitions the following: "a soil almost entirely made up of particles of pure quartz, scarce in nutritive elements, particularly organic matter, but rich in available water due to its great porosity, resulting in much more vegetation than one would expect" (M. Baumer, Noms vernaculaires soudanais, p. 109).
35. For the Arabs the term 'kreb' comprises different grasses.
36. See G. Carvalho and H. Gillet, Catalogue raisonné et commenté des plantes de l'Ennedi, p. 56.
37. An observation confirmed by the study of Hubert Gillet, "Pâturages sahéliens, le ranch de l'ouadi Rimé", in which he remarks that "many grasses possess the property of turning into straw on the spot, while staying upright or slightly inclined for a period and remaining at the animals' disposal throughout the year" (p. 5). He also stresses the fact that "the livestock can content itself with an exclusive diet of hay during the dry season as long as it is watered every day. Thus contrary to what we learn from animal husbandry in temperate regions, one should regard the expanses of hay as true pastures" (p. 17). On the possibility of storing grass see page 85.
38. See pages 14-17.
39. This should be set beside the fact that for a large number of Zaghawa and Bideyat clans, the /teli/ is a sacred tree, /manda/, at whose foot various fertility rituals are performed; see M.-J. Tubiana, Survivances préislamiques, pp. 58-62, 122-123, 158-161, 170.
40. G. Carvalho and H. Gillet, Catalogue, p. 154.
41. See pages 18-25.
42. In "Pâturages sahéliens" H. Gillet mentions trees for browsing, and lists in order of decreasing attraction four species used for grazing in the dry season; they are Maerua crassifolia, Capparis decidua, Acacia senegal and Balanites aegyptiaca (pp. 6 and 97). See page 95, Appetency Index.
43. G. Carvalho and H. Gillet, Catalogue, p. 74.
44. See Michel Baumer, Ecologie et aménagement des pâturages au Kordofan, p. 258 and Noms vernaculaires soudanais, pp. 109-110, and Tothill, Agriculture, pp. 398-399.
45. P. Quezel, Flore et végétation, pp. 64 and 66.

46. The Zaghawa use the term /key/ for small temporary pools as well as the large pools that remain filled with water for part of the year; the pools of water left in its bed by a wadi when it has ceased flowing are called /tubunu/. The Arabic-speaking populations distinguish between: 'fula', 'birke', 'rahad' and 'hafir'; 'rahad' usually designates a larger pool than 'fula', and 'hafir' is used in the Sudan for a pool made by man and the reservoir of a dam.
47. For the first fortnight of December 1965 the vast pool of Um Shedar was reduced to two small pools, but it was everyone's opinion that the rainy season had not been a good one.
48. This is confirmed by a remark of lieutenant-colonel Grossard: "When the pool of Umdur dries up, the people of Tundubay set up a ferik at Rag-rag" (Mission de délimitation, p. 318); Rag-rag is located in the southeast of Tundubay and Umdur.
49. An attempt has been made to equip some of the bore wells with a metal water-wheel operated by means of a handle. We saw this system at Hiri-ba in 1956. It happened that a woman standing on the edge to draw water (this was the only possible way to do it) accidentally fell into the well carrying the handle; the woman was brought up to the surface but the handle remained at the bottom. For a time those using the well turned the wheel by hand, then they returned to the former method of drawing water by rope and bucket.
50. Their case is not unusual: "Among the Mahamid [nomadic Arabs in the Mortcha] the Shigerat group is 'baggara' and the Awlad Jenub group 'albala'. Certain groups, like that of the Ouled Zed, are both" (Capitaine Courtecuisse, "Les Arabes Mahamid", p. 30).
51. The population figures, also provided by tax returns, are always lower than the true figures, but proportionally less so than the figures given for the flocks and herds.
52. The proximity of a vast expanse of desert means that a greater number of camels and sheep can go much further away, consequently leaving more food for larger herds of cattle in the neighbourhood of the village itself.
53. When an animal is abnormal, it is disposed of as quickly as possible. Thus a sheep with five legs or a lamb with two heads would be buried. Such a birth is a bad omen; it means that the chief will die or that an epidemic will carry off many people.
54. See in J. Chapelle, Nomades noirs du Sahara, pp. 181-185, the list of the different species of camels to be found from the Tibesti to the Bahr-el-Ghazal.
55. See A. Le Rouvreur, Sahariens et Sahéliens du Tchad, pp. 302 and 333.
56. According to Zakaria Fodul the female has no horns.
57. This information concurs with the remarks of Sheikh El-Tounsy: "When I was in the Dar-Fur, I admired the slenderness and gracefulness of the sultan's horses ... They informed me that they fed these animals with green fodder consisting of wild grasses from the vicinity of Mount Kouçah, north of Tendelty, and that the lightness and sleek slender profile of the horses was preserved by giving them in addition a rather thick mass of ground doukhn [millet] mixed with honey. Twice a day, added the Sâis, morning and evening, four cupped handfuls of doukhn are kneaded for them; twice a day too they are thrown a few handfuls of the grasses gathered in the vicinity of Mount Kouçah. Furthermore each morning they are given fresh milk to drink. Through this diet the horses gain strength and looks, and remain

graceful and slender, as you can see" (El-Tounsy, Voyage au Ouadây, p.446).

Giving milk to horses is not peculiar to this region; this is illustrated by a piece of information reported by Antoine d'Abbadie. A pilgrim going to Mecca from a country which he called 'al Jaw' was met by d'Abbadie at Adwa in Ethiopia on the 29th April 1842: "I left my home for 'Fas' (Fez) and then for 'Tangiers', where I took a ship to go to Mecca by Egypt ... My country is sandy, has many wells and no rivers, has barley which is eaten by its many horses, and wheat. The horses drink the milk of the camels that abound. The largest town is Walata" (Antoine d'Abbadie, Géographie de l'Ethiopie. Ce que j'ai entendu, faisant suite à ce que j'ai vu, Paris, Mesnil, 1890, pp. 57-58).

58. Recorded in the Sudan; see C. J. Lethem, Colloquial Arabic, who translates 'rifai' as "trotting donkey". In the Chad, G. Trenga notes it as 'rufai', in Le Bura-Mabang du Ouadaï. At Abbéché it was noted as 'rifāy', "tall white donkey" by A. Roth-Laly.

59. A systematic study of animal transfers conducted in a certain number of villages might produce interesting results. It would be useful to know the origin of an animal and the various changes of ownership — this would introduce a further crosscheck to the knowledge of the kinship system. It would also be important to distinguish between the owner of the animal, in other words the person who disposes of it, and the person who enjoys its use. It has been possible to make a start to this work, though it leads to research which is outside the scope of this book.

60. We have not collected any explanations of why a particular clan uses a particular mark; but one soon learns that related clans use similar marks. This chapter is not concerned with a list of marks and families of marks.

61. Talal Asad, "Seasonal movements of the Kababish" and The Kababish Arabs. See also Ian Cunnison, Baggara Arabs, who in the case of the Humr of south Kordofan gives a good description of the cattle's movements related to the resources of water and grass, the type of soils and the absence or presence of flies (pp. 13-22), together with a perceptive analysis of the relations between man and beast (pp. 28-41).

As regards the Chad, an article of Jean-Paul Gilg reviews the different kinds of movements linked to animal husbandry in the Chad Basin, leaving aside the region with which we are concerned here. A map gives the routes of the Kreda, Kecherda and Arabic tribes (J. P. Gilg, "Mobilité pastorale au Tchad occidental et central"); see too J. Chapelle, opus cit., pp. 116, 137, 179-191, and A. Le Rouvreur, 1962 (see Index under: "berger", "élevage", "éleveur", "nomade", "pâturage", "puits", "transhumance", "troupeau", etc.) and also Le Rouvreur, 1971.

See too the special issue of Etudes Rurales devoted to "Terroirs africains et malgaches" (Janv.-Sept. 1970) in which the articles by Edmond Bernus ("Espace géographique et champs sociaux chez les Touareg Illabakan", pp. 46-64) and Henri Barral ("Utilisation de l'espace et peuplement autour de la mare de Bangao, Haute-Volta", pp. 65-84) are of particular interest to us.

62. In his study of the Bedouins of Cyrenaica, E. L. Peters remarks: "fifty is a substantial herd, and a hundred very big ... a poor man or a young man who has not yet had the time to accumulate wealth might possess four or five ..." (The sociology of the Bedouins of Cyrenaica, p. 34).

63. At the beginning of December 1965, at the pool of Um Shedar, we met many herds of cows and goats belonging to some /kaytiŋa/ of Hashaba and people of Bui, Um Beiri and Anka; at that time of year there were not yet many camels. We also saw many travellers among whom a group of Uňay (people from the dar Artaj) which were travelling to the market of Mellit in the dar Berti, to sell goat hair (used for making carpets); a group of seven "Arabs" from Um Kaddada (east of El Fasher) were looking for two of their camels reportedly stolen by Bideyat from Musbat.
64. Apparently /niuw/ does not designate a specific 'goz' but, in this region, all the northern 'goz' as a whole. It would appear to be the equivalent of 'jizu'. We were also given the Arabic term 'safil' as a synonym (see Hillelson: 'safil', "north").
65. This Arabic title designates the head of an administrative district ('omodiya') comprising several villages.
66. See Barbour, The Republic of the Sudan, Fig. 58. The same author refers also to a seasonal movement of the Midob, during the period between November and February, on one side towards the wadi Hawar and on the other beyond the frontier, to the south of the Ennedi, which seems far too extensive to us. On the other hand the journeys of the Kababish towards the 'jizu' pastures, where they meet some Zaghawa, are correctly represented, although they are slightly too extensive when they go right into the Ennedi.
67. We have not been able to carry on our research since our first field-trip of 1956-57.
68. The 'aryal' is the Soemmering's gazelle. These grazing grounds correspond to what the Zaghawa of the Sudan call /sérí/, Ar. 'jizu'.
69. In Europe the calf is first put to suckle, then withdrawn when the milkmaid considers that it has had enough to drink. It is at this point that she draws the extra milk for family use. Among the Zaghawa the woman pulls one or two of the teats while the calf suckles another; for if the cow does not feel her offspring suckling, it withholds its milk.
70. To the suggestion that the dead calf's skin might be stuffed with straw and its mother might thus be tricked into believing that it still has its calf, the Zaghawa reply: "this would be pointless since the calf must suckle for a few minutes for the cow to give its milk".
71. Zaghawa informants give the term 'garus' as an Arabic equivalent (see Hillelson: 'gāris' which means in western Sudan "sour camel's milk kept fresh in a skin", p. 190).
72. However a very thirsty man is given half milk, half water.
73. The Kobé say that among the Tuer of the Sudan men milk the cows. In fact this is without foundation. It is just a way of disparaging their neighbours.
74. To make belts they prepare a number of twisted threads and stretch them horizontally by tying their ends at two other threads stretched between two pairs of small wooden pegs. They then use a wooden needle for weaving.
75. See Hillelson: 'murāh', "flock" and also M. Baumer: 'morah', "animal enclosure, of permanent or temporary construction, more elaborate than that of the 'fariq' or the 'zariba'" (Noms vernaculaires, p. 113).
76. However an informant adds: "in the Kobé it is the blacksmiths who keep the herds and flocks of other people for a wage". We were unable to check this.
77. Millet is directly made into flour without steeping the seeds for two or three days as is usually done.

78. This practice has been reported among the Tuareg. In northern Ethiopia the beestings are cooked and eaten.
79. See G.-J. Lethem, p. 260: 'marara', "bile".
80. Sometimes the inside of the heart is consulted to find out the intentions of the person who slaughtered the animal. They are regarded as bad if coagulated blood is found inside, good in the opposite case. The informant adds: "little importance is attached to it, it's rather fun".
81. Tisserand, "Rapport au Maréchal Vaillant", Moniteur universel, 8th April 1868, p. 494, col. 1 (quoted by Littré under "Transhumance").
82. Hubert Gillet, "Pâturages Sahéliens", pp. 95-107, and Pierre Bourreil, "A herbivore appetency evaluation", pp. 23-25. See page 95 above.
83. This does not apply to regions where the gathered grass par excellence is Cenchrus biflorus, for instance in Air.
84. This does not apply to regions totally lacking in trees, where the Calotropis provides wood, as in Borkou. M.Baumer reminds us, in a recent letter, that the Calotropis grows on already impoverished soils.
85. See Chapter 3.
86. See note 37 above.
87. See Chapter 2.
88. See mainly pages 18-25 above.
89. Research along these lines has been conducted in the Niger by C. Raynaut, Quelques données de l'horticulture, p. 31.
90. Without precluding the introduction of non-African types. For instance an American variety of watermelons has been successfully acclimatised in Dar-Fur, where it is known under the name of "Rothmans", by our Zaghawa friend, Mahmud Beshir Jamma.
91. This incident has already been described in note 49.
92. The Zaghawa have camels; they might also try to replace the mule by a very robust local breed of donkey, the 'rifāi'.
93. See R. Capot-Rey, Le Sahara Français, pp. 320-321. According to this author the animal-operated well, worked by a donkey, was introduced some years ago to the Borkou by Libyans (Borkou et Ounianga, p. 111). Ch. Le Coeur mentions such wells, worked by cows, among the Daza (see 'yige doroso' in Dict. ethnographique teda, p. 196). More recently M. Le Coeur and C. Baroin have observed this type of well, worked by a donkey, a camel or a cow, among the Daza and Azza of the Niger Republic.
94. The shadoof (Arabic 'šādūf') is made of a pole with a bucket at one end and a counterpoise at the other. It is described as follows in Tothill, Agriculture in the Sudan, p. 952: "a hand-operated water-lifting device suited for watering plots of vegetables. It is on the seesaw model with the water container counterbalanced with a lump of clay". According to Allan and Smith, it is "a very primitive but quite effective means of lifting water by man-power through a limited height, usually from a pool, canal or river. It is cheap and simple to construct and maintain. It consists essentially of two wooden posts or pillars of dried mud, supporting a cross-bar on which is pivoted a long wooden lever. To the shorter end of the lever is fixed a stone or ball of dried mud; this acts as a counterpoise to a rod or rope and dipper attached to a longer arm. Below this end is the inlet channel from which the water is to be lifted. The rod is seized high up and pulled down until the dipper enters the water. The full dipper is then allowed to rise, pulled up by the counter-

weight, until it reaches the level of the upper channel, into which it is emptied by a sideways tilt. The dipper may be a bag of leather, but the 4-gallon petrol tin is commonly used. Lifts of up to 3 metres can be obtained, but the greater the lift the fewer the strokes per minute. At a lift of 20 m a shadūf worked by one man would have an output of about 3-5 m^3 per hour or, say, 24-30 m^3 per day..." (Tothill, ed., Agriculture in the Sudan, p. 631).

95. R. Capot-Rey, Le Sahara Français, pp. 322-323. See also C. Raynaut, op. cit., p. 18 which gives a very accurate description and Ch. Le Coeur, op. cit., entry: 'yoba', together with Capot-Rey, Borkou, p. 111. This kind of well is also to be found among the Berti mentioned above.
96. A. Leroi-Gourhan, L'Homme et la Matière, Evolutions et Techniques, Paris, Albin Michel, 1943, p. 98.
97. See M.-J. Tubiana, Survivances, pp. 55-56.
98. One might also suggest the production of powdered meat, by crushing bones and meat — as was done in Kenya in the years 1959-62 in order to feed populations suffering from drought and as a means of saving something from animals that would in any case not survive (oral information given by A. H. Jacobs). It should be noted that the Zaghawa and other Sudanese when using dried meat (Ar. 'šarmut') crush it to powder in a mortar before using it as an ingredient of sauces.
99. Oral teaching should make extensive use of the vernaculars. One might also consider the possibility of showing Zaghawa-speaking educational films.
100. In 1970 a number of Zaghawa leaders, with traditional backgrounds or a modern education, presented their problems to us and asked our opinion on the questions they were debating. We held long discussions with them examining point by point possible solutions to their difficulties. The discussions took up a number of sessions separated by a few days' interval, thus giving us time to think, to introduce new problems and to work out the means of putting into effect the improvements on which agreement was reached.

    This interesting experience, which we underwent at the behest of our Zaghawa friends, convinced us that from whatever way one looks at it, as an ethnologist, agronomist, economist, government official, technical expert and so on, one must rid oneself of one's prejudices as regards the nomadic way of life, if one genuinely wishes to help in a really practical way. It must be noted that until now the majority of administrative bodies entrusted with the task of settling the problems of nomadic societies, are recruited among sedentary city-dwellers; hence perhaps their failures when confronted with a task for which they are unprepared. These are the exchanges and considerations from which this paper originated.
101. The years 1972-73 were extremely bad, as everybody now knows. Our young friend Ishaq Adam Beshir wrote from Khartoum, just at the end of the dry season in 1973:

    "As for the last two years there is complete drought in Zakhawa only so most of the tribe tried to emigrate to the south of Darfour where there is a plenty of rain. The drought reaches max. this year. Suppose you have heard of it. It is extension to the drought happened to Western Africa. There is a suggestion from the Government to locate all the tribe in a place called Hofrata el Nahas, a place at the southern boundaries of Darfour and some [of] the people agreed but others refusing I don't know what will be the final result but soon it will be clear after the rainy season". (Khartoum, 9/7/73).

SELECTED BIBLIOGRAPHY

Asad, Talal
    1964. "Seasonal movements of the Kababish Arabs of northern Kordofan", Sudan Notes and Records, 45: 48-58.
    1970. The Kababish Arabs, London, C. Hurst.

Barbour, K. M.
    1964. The Republic of the Sudan, University of London Press, 2nd edition.

Baumer, Michel
    1968a.Ecologie et aménagement des pâturages au Kordofan, Thèse de Doctorat d'Etat, Faculté des Sciences, Université de Montpellier (mimeog.).
    1968b.Les Noms vernaculaires kordofanais utiles à l'écologiste, Thèse complémentaire, Faculté des Sciences, Université de Montpellier (mimeog.).
    1975. Noms vernaculaires soudanais utiles à l'écologiste, Paris, Editions du C.N.R.S. (A revised version of the preceding thesis).

Bernus, Edmond
    1967. "Cueillette et exploitation des ressources spontanées du Sahel Nigérien par les Kel Tamasheq": Cahiers O.R.S.T.O.M., Sciences Humaines IV series, 1: 31-52.

Bourreil, Pierre
    1968. "A herbivore appetency evaluation index for certain Darfur Province plants (Sudan)." Dossiers of the R.C.P. 45 (4): 23-25.

Capot-Rey, Robert
    1953. Le Sahara Français, Presses Universitaires de France, Paris.
    1961. Borkou et Ounianga, Etude de géographie régionale, Mémoire de l'Institut de Recherches Sahariennes, 5, Alger.

Chalmel (Capitaine)
    1931. "Notice sur les Bideyat", Bull. Sté de Recherches Congolaises, 15: 33-91.

Chapelle, Jean
    1957. Nomades noirs du Sahara, Paris, Plon.

Courtecuisse, Louis
    1971. "Les Arabes Mahamid du district de Biltine", in Quelques populations de la République du Tchad, Paris. (Recherches et documents du C. H. E.A.M., III.)

Cunnison, Ian
    1964. Baggara Arabs, Oxford, Clarendon Press.

Gast, Marceau
- 1972. "Céréales et pseudo-céréales de cueillette du Sahara central (Ahaggar)": Journal d'Agriculture Tropicale et de Botanique appliquée, 19 (Janv.-Fevr.): 50-58.

Gilg, Jean-Paul
- 1963. "Mobilité pastorale au Tchad occidental et central", Cahiers d'Etudes Africaines, 3 (12): 491-510.

Gillet, Hubert
- 1960. Catalogue raisonné et commenté des plantes de l'Ennedi, Office antiacridien, Bulletin hors série, Janv.-Août. (in collaboration with G. Carvalho).
- 1961. "Pâturages sahéliens, le ranch de l'ouadi Rimé", Journal d'Agriculture Tropicale et de Botanique appliquée, 8 (Oct.-Nov.): 3-210.

Grossard (Lieutenant-colonel)
- 1925. Mission de délimitation de l'Afrique Equatoriale Française et du Soudan anglo-égyptien, Paris, Larose.

Hillelson, S.
- 1930. Sudan Arabic, English-Arabic vocabulary, London, published by the Sudan Government, 2nd edition.

Lebon, J. G.
- 1965. Land use in Sudan, The world land use survey, 4, Sir Dudley Stamp.

Le Coeur, Charles
- 1950. Dictionnaire Ethnographique Téda, Paris, Larose.

Le Rouvreur, Albert
- 1962. Sahariens et Sahéliens du Tchad, Paris, Berger-Levrault. Index published separately in 1968 (Dossier No. 3 of the R.C.P. 45).
- 1971. "Une saison sèche en Ennedi (1949-1950)", Etudes Rurales, 42: 172-177, 1 map.

Lethem, G. J.
- 1920. Colloquial Arabic. Shuwa dialect of Bornu, Nigeria and of the region of Lake Chad, London, Crown Agents for the Colonies.

MacMichael, H. A.
- 1912. "Notes on the Zaghawa and the people of Gebel Midob": J.R.A.I., 42: 288-335.
- 1922. A history of the Arabs in the Sudan and some account of the people who preceded them and of the tribes inhabiting Darfur, 2 vols., Cambridge University Press.

MacRamsay, C.
- 1958. The forest ecology of central Sudan, Agricultural Publications Committee, Khartum.

Maurizio, (Dr.) A.
- 1932. Histoire de l'Alimentation végétale depuis la préhistoire jusqu'à nos jours, Paris, Payot.

Monod, Th. (editor)
- 1975. Pastoralism in Tropical Africa, London, O.U.P.

Peters, E. L.
- 1951. The sociology of the Beduins of Cyrenaica, unpublished thesis, Lincoln College, Oxford.

Quezel, Pierre
   1969. Flore et végétation des plateaux du Darfur nord-occidental et du Jebel Gourgeil, Dossier No. 5 of the R.C.P. 45.

Raynaut, C.
   1969. Quelques données de l'horticulture dans la Vallée de Maradi, Etudes Nigériennes, 26.

Roth-Laly, Arlette
   1969-72. Lexique des parlers arabes tchado-soudanais. An Arabic-English-French Lexicon of the dialects spoken in the Chad-Sudan area, Paris, C.N.R.S., published in 4 parts.

El-Tounsy, Mohammed Ibn-Omar
   1845. Voyage au Darfour, translated from Arabic by Dr. Perron, Paris, B. Duprat.
   1851. Voyage au Ouadây, translated from Arabic by Dr. Perron, Paris, B. Duprat.

Trenga, G.
   1947. Le Bura-Mabang du Ouadaï, Paris, Institut d'Ethnologie.

Tothill, J. D. (editor)
   1954. Agriculture in the Sudan, London, O.U.P., 3rd edition.

Tubiana, Joseph
   1960. "La mission du Centre National de la Recherche Scientifique aux confins du Tchad", Cahiers d'Etudes Africaines, 1: 115-120.
   1961. "Moyens et méthodes d'une ethnologie historique de l'Afrique orientale", Cahiers d'Etudes Africaines, II, 1(5): 5-11.
   1963. "Note sur la langue des Zaghawa": Travaux du XXVe Congrès International des Orientalistes, Moscou, 9-16 août 1960, Volume V, Moscou, pp. 614-619.

Tubiana, Marie-José
   1960. "Un rite de vie: le sacrifice d'une bête pleine chez les Zaghawa Kobé du Ouaddaï": Journal de Psychologie, 3: 291-310.
   1961. "Le marché de Hili-ba: moutons, mil, sel et contrebande": Cahiers d'Etudes Africaines, 6: 196-243.
   1964. Survivances préislamiques en pays zaghawa, Paris, Institut d'Ethnologie, LXVII.
   1975. "Exogamie clanique et Islam: l'exemple Kobé": L'Homme, XV(3-4): 67-81.

Tubiana, Marie-José and Joseph Tubiana
   1962. Contes zaghawa, Paris, Quatre Jeudis.
   1967. "Mission au Darfour": L'Homme, 7(1): 89-96. (English translation: "Field work in Darfur", Dossiers of the R.C.P. 45, 1968, pp. 1-16.)

INDEX

The following abbreviations are used in this index:

App. : Appendix
Ar. : Arabic
pop. : population
var. : variant
vill. : village

Place names and peoples' names are in capital letters except when transcribed. Zaghawa words are transcribed between oblique bars. Arabic and Teda words are transcribed between single quotation marks.

/abbo/: prince 40
'absabe' or 'abu sābi' (Ar.), see /bóù/
'abu hirēbat' (Ar.), App. Albizzia
'abu jigra' (Ar.), App. Brachiaria sp.
'abu nigēgēra' (Ar.), App. Alysicarpus
'abu raxis' (Ar.), App. Andropogon gayanus
'abu wīhīdu' (Ar.), App. Grewia
Acacia Faidherbia albida, see /teli/
Acacia mellifera 35
Acacia raddiana, see /kedira/
Acacia scorpioides, see /bírgè/
Acacia senegal, see /túè̂/
Acacia seyal, see /musumará/
ADERA (well) 55
/ā́dī̆/ (Ar. 'berbere'?): cracked soil 34
ADRAR 88, 89
'afan al xadīm' (Ar.), App. Chloris gayana
'agēg al bagar' (Ar.), App. Aristida stipoides
/ágǐ/, see /kǐ̀/
agriculture 6, 85-86
'agul' (Ar.): Fagonia cretica 36
AHAGGAR 89
/áīgǐ/ (Ar. 'sēf'): dry and hot season 5, 17, 32, 33, 52, 55, 68, 70, 71
AIR 88, n83
/áīrī/ (Ar. 'kreb'; Echinochloa colona ?): a wild grass 15, 34, n11
see also 'kreb'
'akariš' (Ar.), App. Pappophorum
AMBAR (wadi) 53, 68
'am-belele' (Ar.), see /tómsò/
'am-hoy' (Ar.), see /búbù/

AM SUGAT (wadi) 55
'am zerar' (Ar.), see /ɲ́áʀī/ (?)
'andarab', 'andrab' or 'indarāb' (Ar.), see /túrù/
animals
  abnormal n53
  acquisition of 46, 48-49
  census of 40, 41, 43, 51
  loaning of 48
  movements of 49-72, n61, n66
  transactions of 8
  see also: cattle, camels, donkeys, goats, horses, sheep, flocks and herds, livestock
ANKA (vill., wadi and wells) 26, 32, 34, 38, 41, 49, 50, 51-52, 54, 68, 75, 87, n63
/áŋ́ará/: Hibiscus sabdariffa 24
/aŋu/ (clan) 47
ARABIC TRIBES 27, 46, 52, n61
arabicisation 1-2
'arad' (Ar.), App. Albizzia
arboriculture 86-87
'ardeb' or 'aradēb' (Ar.), see /médèr/
Aristida: wild grasses 34
Aristida papposa 36
'arkala' (Ar.), App. Ipomoea repens
AR - KOILA (well) 38
ARKURI (well) 54
ARTAJ (Zaghawa) 43, 49, 53, 70, n63
Artemisia judaica, see 'ediseru'
'asal al tēr' (Ar.), App. Leucas
Asclepiadaceae, see /kórfú/

111

'askanit' (Ar.), see /nógò/
'ašrut' (Ar.), App. Indigofera, Tephrosia nubica
ATI 42
'atia' 42
ATTAWIA (well) 53
/ayar/: ripe fruit of /mádɪ̄/ 22
AZZA (blacksmiths among Daza) 27, n16, n93
/bà/ (Ar. 'bir'): permanent timbered well 37
'bābanus' (Ar.), App. Dalbergia melanoxylon
BA - HAY (well and wadi) 3, 69, 70, 71
BAKAORÉ (vill.) 7
Balanites aegyptiaca, see /gɪ̃è/
BA - MESHI (well) 54
BA - MINA (well and wadi) 3, 68, 69, 70
'banu', see 'bonu'
barter 8, 28, 29
BA - SAO (wells and dam) 37, 68, 69, 70, 90
BASSO 3
'basturma': a kind of smoked meat 91
/bátɪ̄/: drinking-trough 38
'battix' or 'bitēx' (Ar.), see /órù/
'baxšem', 'baxšamay' (Ar.), see /sóŋò/
beer 17, 22
BEIRI (dar) 43
BEJA 43
'belile', see 'am-belele'
'berbere' (Ar.), see /ãdɪ̄/
/bèRɪ́/ 1, 27
BERTI n95
BIDEYAT 1, 2, 10, 16, 17, 46, 49, 52, 53, 69, 70, 77, n1, n32, n39, n63
'bigel' (Ar.), App. Blepharis linariifolia
BILTINE 41
BI - MARA (wells) 55
/bɪ́nɪ̄/ (Ar. 'bonu' or 'bolu'; Pennisetum tiphoideum ?): a wild grass 16
'bir' (Ar.), see /bà/
BIR ATRUN (Z. teDi-bà): wells 53
BIR EN (wells) 54
BIR FOKHMA (wells) 52
/bɪ́rgè/, /bɪ́rgéRà/ (Ar. 'garad'; Acacia scorpioides) 25, 35
/bìrɪ̀/: light brown 42
/biriaRa/ (Bideyat clan, part of which live in dar Tuer) 53
BIR IZ EL-KHADIM (wells) 52
'birke' (Ar.): pool n46
BIR - TAWIL (wells) 38
BIR TOMUR (wells) 52
BIR TUNDUB (wells) 54
'bišari': race of camels 43
'biyēd' (Ar.), App. Commelina
blacksmiths 5, 6, 7, 21, 45, 50, 79, 80, 92, 93, n16, n76
   see also AZZA, 'duudi'
/bógóù/, see /ɪ̀rsǎsi/

'bonu', 'bolu' or 'banu' (Ar.), see /bɪ́nɪ̄/ App. Eragrostis
Boraginaceae, see /túrù/
BORKOU 41, n84, n93
borrowing, see animals (loaning of)
/boru deni/: fresh butter 79
/boru ergi/: fat inside the hump of a camel 80
/boru hamu/: clarified butter 79
/boru soru/ 79
Boscia senegalensis, see /mádɪ̄/
/bóù/ (Ar. 'absabe'; Dactyloctenium aegyptiacum): a wild grass 14-17, 34, 84, 86, n31
BOW - BA (wells) 55, 69
Brachiaria deflexa: a wild grass 34
bridewealth 9, 28, 41, 47
browsing, see grazing
/búbù/ (Ar. 'am-hoy' or 'kwoinkwoin'; Eragrostis pilosa): a wild grass 16, 17
BUI (goz) 51, n63
Burseraceae, see /tógórò/
butter 62, 79-80

Caesalpiniaceae, see /médèr/, /káwán/
Calotropis procera, see /kórfú/
camels 39, 40, 41, 43-44, 51, 52, 75-76, n7, n54
   camel-breeders 39
   camel-load 44
   colour of hair 43
   markings 47
   morphological features 43
   size of camel herds 50-51, n62
   thieving 8, 47, n63
   see also flocks and herds movements, milk, water
capitalism 7
Capparidaceae, see /mádɪ̄/, /námár/, /núr/
Capparis decidua, Capparis sodada, see /námár/
cashcrop 8, 26
Cassia tora, see /káwán/
cattle 7, 40, 41, 42, 43, 50, 51, 73-74, n7
   cattle-breeders 7, 39, 41
   colour of the coat 42
   morphological features 42
   names 42, 73
   sale of 7
   size of cattle herds 50
   see also flocks and herds, seasonal movements, milk, water
Cenchrus biflorus, see /nógò/
cereals, wild cereals 14-18, n9, n11, n15, n31
   see also millet
CHAD 4, 6, 7, 8, 10, 13, 26-29, 39-41, 43, 44, 49-50, 72, 76, 78, n1
chief, chiefdom 9, 10, 93

clan 9-10, n16, n39, n60
    clan's brands 46, 47, 78, n60
    clan's prohibitions n16
    clan's rights 17
climate 32-33, 83
Coccinia grandis, see /túdù/
Colocynthis citrullus, see /órù/
Colocynthis vulgaris 24
Colonial powers (Great Britain and France) 2, 10, 41, 88
Commiphora africana, see /tógórò/
Corchorus olitorius, see /múlúkíé/
Cordia gharaf, see /túrù/
Cordia Rothii, see /túrù/
crafts and small industries 7, 91-92, n63, n74
Crotalaria thebaica, see 'nataš'
Cruciferae, see /sundu/
Cucurbitaceae, see /oru/, /túdù/
cultivation 3, 6
Cyperaceae, see /nógù/
Cyperus rotundus, see /nógù/

DABAY (wells) 69
/dábó/ (Ar. 'šitā'): dry and cold season 4, 32, 33, 49, 52, 53, 68, 70, 71
Dactyloctenium aegyptiacum, see /bóù/
DAGAL (pop.) 27
'dalu', 'delu' (Ar.) 88-89
  see also /kwòí/
dams 37, 90, n46
'dannab al nāga' (Ar.), App. Schoenefeldia gracilis
'dannab al fallu' (Ar.), App. Setaria pallidefusca
'darat' (Ar.), see /tàrbà/
'darat al baggara' (Ar.), App. Portulaca
DAR - FUR (sultanate and province) 2, 28, 36, 43, 49
DARMA (wells and village) 54, 69
dates, date palms 7, 86-87
DAZA 3, 27, 46, n93
DÉMI (rock-salt mine) 48
'derma' (Ar.): Indigofera bracteolata 36
desert 2, 36, 94, n52
/dìdì/ (Ar. 'warwar'): stirrer 79
diet 13, 15, 16, 17, 19, 20, 21, 22, 23, 24, 25, 26, 78-81, 87
  of the herdsmen 76
  during Ramadan 20, n31
'difra' (Ar.): a wild grass n9
DILDILA (wells) 69
/diŋe/: a kind of dough 15
DIRONG (Zaghawa) 17, 40
DISA (vill.) 7
DI - SHERI (wells) 55
donkeys 40, 41, 43, 45, 51, n58
  size of 50
DOR (vill. and wells) 7, 24, 49, 50, 51, 52, 53, 68, 87

'drese' or 'daresa' (Ar.), see /tárà/
drought 43, 82, 83, 93-94, n98, n101
/dùgún/: flail 15
DUGU - RAY (place name) 46, 54, 68
'duudi' (blacksmiths among Téda) 27

Echinochloa colona, see /áírí/, /jígíRí/
economic organization 5-8, 26-28, 81
'ediseru'(Teda; Ar. 'šii'; Artemisia judaica) 18
education 92-93, n99
/ègé/ (Ar. 'kreb'): a wild grass 16, 34
/ègìmè-bur/: young locust 26
EGYPT 8, 76, 89
/éígò/: tar 17
/eigo-biRi/ (Ar. 'kilwe' or 'xirwe'; Ricinus communis): castor-oil plant n15
/elbira/ (clan) 47
EL FASHER 49, 69
ENNEDI 3, 4, 36, 41, 49, n32, n66
environment 2-5, 9
Eragrostis: wild grasses
  E. aspera n13
  E. cilianensis, see /mine/
  E. pilosa, see /búbù/
ERDEB (wadi) 68, 69, 70, 71
EREGAT 49
/erfe/: brand of clan 47, 78
ER - SHEHARI (wells) 55
ETHIOPIA 91, n73
Euphorbiaceae, see /eigo-biRi/
export 8, 26, 91
'ēyer' (Ar.), App. Momordica balsamina

FADA (vill.) n32
Fagonia cretica, see 'agul'
faki: a learned man 10, 45
famine 83, 93-94
FATTA BORNU (vill.) 86
'ferik' (Ar.): temporary hamlet 45, 46, 52, 54, 69, n75; see also /guli/
FEZZAN 88, 89
'fileya' (Ar.), see /ósù súlí/
flocks and herds
  seasonal movements of 49-72, n66
    at Anka 51-52, 58-59
    in the dar Artaj 53, 62-63
    at Dor 52, 60-61
    in the dar Gala 54-55, 64-65, 69
    in the dar Kobé 66-67, 69-71
    in the region of Oru-ba 66-67, 71-72
    in the dar Tuer 54, 62-63
    in the region of Tundubay 66-67, 71
  size and composition of 40-42, 43, 51
  see also livestock
FOKHMA (wadi) 53
FORAWIYA (vill., wadi and wells) 32, 44, 46, 54, 55, 68, 69, 75

FROLINAT: Front de Libération Nationale du Tchad 2
fruits 87, n90
    wild fruits 13, 18-25, 86, n9, n18, n20
'fula' (Ar.): pool n46
FUR 3, 43, n9;
    see also DAR - FUR
FURNUNG (dar) 87

GADIR (wadi) 68, 70
'gafal' (Ar.), see /tógórò/
GALA (Zaghawa) 43, 49, 50, 51, 54-55, 69, 70
'garajub' (Ar.), App. Oldenlandia
'garad' or 'garat' (Ar.), see /bírgè/
GARDAI (wadi) 55, 68
'garen' (Ar.), App. Mollugo cyclocarpa, Monsonia senegalensis
'garis' var. 'garus' (Ar.): curdled milk n71
gathering
    food gathering 5, 13-28, 84, 85, 86
    ritual and 18
'gau' (Ar.), App. Aristida adscensionis, Aristida funiculata, Aristida mutabilis, Aristida rhinochloa
/génè/, /génèRà/ (Ar. 'himed'; Sclerocarya birroea) 21
/genigergeRa/ (clan) 47, 69
GERDI (wells) 69
'gerēdān' (Ar.), App. Grewia tenax (?)
'gerēgerām' (Ar.), App. Grewia
/gèrèj/: a breed of sheep 44
GEZIRA 7, 49
'giddem' or 'gudēm' (Ar.), see /ñárî/
/gíê/, /géyRà/ (Ar. 'hejlij'; Balanites aegyptiaca) 21, 35, 85, 86, n9, n42
GIMIR 3, 70, 71
goats 40, 41, 42, 43, 51, 75, n7
    names of 42
    size of herds 50
    see also milk, water
/godu/: wooden dish 44
/gógúrù/: the stone of the /gíê/ 21
/góróù/: the kernel of the /gíê/ 21
'goz' (Ar.): a sandy expanse 34, 46, 52, 54, 68, 71, n34, n64; see also /šige/
GOZ EL ARAB 51
GOZ EL HARR 51, 52
GOZ KHARRA 52
GOZ KHURR 52
GOZ KORU-HAY 52
GOZ LEBAN 51
GOZ NAY 52
GOZ NIUW 53
Gramineae, see /áìrí/, /bíní/, /bóù/, /búbù/, /jígírí/, /miné/, /nógò/, /sábà/

grazing 3, 5, 27, 35-36, 46, 84-85, 93, n37 n42
    degree of appetency 84
    property of pastures 46
    selectivity 36, 84
    storage of hay 85
    various types of pastures 84
    see also 'jizu'
Grewia flavescens, see /gúgúr/
Grewia populifolia, see /ñárí/
Grewia villosa, see /kòrfù/
GREYGI (wadi) 67
/gú/: porridge 15, 34, n11
'gubēn' (Ar.), App. Solanum
'gubeš' (Ar.), App. Guiera senegalensis
/gúgúr/, /gúgúrdà/ (Ar. 'kabayŋa'; Grewia flavescens) 19, 35
/guli/ (Ar. 'ferik'): a temporary hamlet 52 (Ar. 'zeriba'): a thorny enclosure 73
gum arabic 26, 86, n23
GUMGUM (goz) 52
GUR'AN (pop.) 36
GURUF (Zaghawa) 3, 17, 24, 34, 40, 50, 51, 74
'guttub' (Ar.): Tribulus longipetalus 36
'gutun' (Ar.): the cotton-plant, App. Gossypium
/gwa dei/: crow's foot 47
/gyé/ or /ídí gyé/ (Ar. 'xartí'): rainy season 4, 5, 32, 51, 53, 68, 70, 71
Gynandropsis gynandra, see /ŋáî/

'hafir' (Ar.): reservoir n46
HAJER JUWA 70
/hámáréi/: reddish 42
'hantut' (Ar.), App. Ipomoea
'haraz' (Ar.), see /teli/
/hárí/: dappled 42
/hasaniya/: race of goats 42
HASHABA n63
'haskanît' (Ar.), App. Cenchrus biflorus, Cenchrus ciliaris; 'haskanît naɛam', App. Cenchrus prieurii. The Zaghawa usually pronounce 'askanit', see /nógò/
'hašāb' (Ar.), App. Acacia senegal
HAWAR (wadi) 55, 68, 69, n66
HAWASH (wadi) 3
HA - WISHA (jebel) 53
hay, see grazing
HA - YEI (jebel) 38
'hejlij' (Ar.), see /gíê/
herdsmen 13, 22, 23, 39, 46, 72-78, 92
    contact between 46, 53, 69
    paid workers 76-78
    unpaid member of the family 76-78
Hibiscus esculentus, see /nyaRi/

Hibiscus sabdariffa, see /áŋárá/, /kerkere/ or /kerkedi/
hides and skins 5, 25, 80-81, 92
HILALIA (vill.) 71
'himed' (Ar.), see /génè/
HIRI - BA ('IRIBA' on French maps; /hiʀi-ba/, the cows' well) 7, 32, 38, 40, 41, 88, n3, n49
history 1-2
HOFRAT EN NAHAS 4, 93, n101
horses 17, 23, 27, 40, 41, 43, 44-45, 51, n7, n57
size of 50
horticulture 87
HOWAR (wadi) 3
HUMA - BA (wells) 69
'humēd' (Ar.), App. Sehima ischaemoides
HUMR n61
hunters 5
HURIO - KOILA (wells) 69

/ídí gyé/, see /gyé/
IKLAN 27
/ila digen/ (clan living in the dar Tuer) 53
IMAM (jebel) 52
/imogu/ (clan) 47
import 7
independent states (Sudan and Chad) 2, 37-38
Indigo arenaria, see 'xušĕn'
Indigofera bracteolata, see 'derma'
IRIBA, see HIRI - BA
irrigation 91
/ìrsāsi/ (Ar. 'uršăš') or /bógóù/: short period preceding the rainy season 32, 68, 70
Islam, islamisation 2, 10
Maliki system 10

JAK (jebel) 52
/jígíʀí/ (Ar. 'kreb'; Echinochloa colona ?): a wild grass 15, 34, n11
/jina/: a black cow with a white tail etc... 42
JINNIK (pool) 37, 49, 52
'jizu' (Ar.): a kind of pasture 36, 46, 49, 53, 68, 69, 70, 75, 84, n64, n66, n68
see also /séʀí/
/jórdò/: broom 14
/jornok/: whooping cough 79
/jude/ (clan) 90
'jumēz' (Ar.), App. Ficus

KABABISH 36, 42, 44, 46, 49, 53, 69, 94, n61, n66
/kábárà/, /kábáîʀà/ (Ar. 'nabak'; Ziziphus mauritiana, Ziziphus spina-christi): jujube tree 21, 35, n9
/kābāš/ or /seri/: a breed of sheep 44
'kabayŋa' (Ar.), see /gúgúr/
KABKA (Zaghawa) 4, 34, 37, 40, 49, 70
'kadād' (Ar.), App. Dichrostachys sp.
KAIDA - BA (vill.) 69, 70
KAMARDA (wells) 69
KAMO (vill., wadi and wells) 46, 54, 55, 68, 69
KANEMBU 27
/kara/: sacks 81
/kárákárá/: soil with pebbles 34
/karda/: herdsman 77
/kárkáń/: leaves of Hibiscus sabdariffa 24
KAWAHLA 36
/kawal/, var. /káwán/ (Ar. 'kawal'; Cassia tora) 26; App. Cassia occidentalis
/kaytiŋa/ (clan) n63
KECHERDA n61
/kedira/ (Ar. 'seyal'?; Acacia raddiana) 35
'keldos' (Ar.), App. Papilionaceae
KEREINIK (vill.) 69, 70
/kerkere/ or /kerkedi/ (Hibiscus sabdariffa) 24
/keti/: rope 38
/key/: a pool n46
KEY - BA (wells) 69
KEY - HAY (pool) 37
KHARTUM 44
KHERBAN (jebel) 52
/kǐ/, var. /ágǐ/: hooked stick 47
KIBET 27
/kíè/, /kéîʀà/, (Ar. 'korno'; Ziziphus spina-christi, Ziziphus mauritiana): jujube tree 17, 20, 35, 86, n9, n19
'kilwe' (Ar.), see /eigo-biʀi/
/kíʀǐ/: curved horns... 42
'kitir' (Ar.), see /túẽ/
KOBA (wadi) 35, 38
KOBÉ (Zaghawa) 10, 13, 16, 29, 34, 37, 40, 43, 49, 50, 51, 55, 69-71, 72, 77, 85, 90, n73
/kómùn/: seeds contained in the fruit of Hibiscus sabdariffa 26
/kóndù/: pulp of /kíè/; mixture of /tárà/ flour and /kíè/ flour 17, 20
KORDOFAN 36, 42, 49, n61
/kordofaniya/, /kordofale/ 42
/kòrfù/, /kòrfúʀà/ (Ar. 'tomur el abid'; Grewia villosa) 19, 35
colour of animal coat 42
/kórfú/ (Ar. 'ušar'; Calotropis procera) 25, 84, n84
'korno' or 'karnuya' (Ar.), see /kíè/
KORNOY (vill., wadi and wells) 7, 44, 55, 69, 70, 75
'kreb' or 'kurēb' (Ar.) 15-16, 17, 19, 44, 71, 84, 86, n9, n35; see also /áîʀí/, /ègé/, /jígíʀí/, /sábà/: wild grasses
KREDA n61
KUKA 27
'kulkul' (Ar.), App. Bauhinia rufescens
KULKUL (wadi) 55
/kúná/ (Ar. 'tamada'): water hole 37

115

'kurmut' (Ar.), see /núr/; App. Cadaba
  glandulosa
KUTUM (town) 32, 37, 86, n3
/kwòí/ (Ar. 'dalu'): dipper 37
'kwoinkwoin' (Ar.), see /búbù/

'laɛut' (Ar.), App. Acacia nubica
'layūn' (Ar.), App. Lannea humilis
LIBYA 8, 36, 42, n93
LIL (wadi) 54
livestock 39-45
  'basic herd' notion 40
loaning, see animals
'lusēg' (Ar.), App. Zornia diphylla

MABA 27, n16
/mádì/, /mádírà/ (Ar. 'moxet' or 'muxīd';
  Boscia senegalensis) 17, 22, 35, 86, n9
/madí/: small trough 44
Maerua crassifolia, see /núr/
MAGHREB (wadi) 54, 68
MAHALLAT KWOILA (place name) 53
MAHAMID 3, n50
Maliki system, see Islam
Malvaceae, see /áɲárá/, /nyaʀi/
/mamur/: copper bow 47
/mándà/: spiritual power 14, n39
manpower 7, 78, 92
'marara' (Ar.): a dish 80, n79
MARARIT 3, 27
/márdì/ (Ar. 'naga'): clay soil 34
'mardo' (Ar.), App. Capparis tomentosa
'marhabeb' (Ar.), App. Chrysopogon sp.,
  Cymbopogon nervatus (?)
markets 7, 23, 27, 28, n63
MARRA (jebel) 3
marriage 10
  exogamous marriage 9, 10
  see also bridewealth
MASALIT 27, 49
MATADJÉNÉ (pool) 7, see also key-hay
meat 7, 80-81, 91
  dried and smoked 91
  powdered n98
/médèr/, /médèrdà/ (Ar. 'ardeb'; Tamarindus
  indica): the tamarind 25, n8
medicine
  medicinal plants 25, 26
  abortive remedy 25
  chicken pox 25
  eye-sores 25
  haemorrhages 25
  medical treatment 76, 79, 80
MELLIT (town) 37, n63
'merise' (Ar.): beer 17, 22
'merx' (Ar.), App. Leptadenia pyrotechnica
METEL - KORU (dam) 37, 90
MIDOB 36, n66
migration 93, 94, n101

milk 6, 48, 73, 76, 78, 91, n71, n72
milking camels 76
milking cattle 73, 74, n69, n70, n73
milking sheep and goats 75
millet
  bulrush millet (/bàgà/, Ar. 'duxn') 6, 7,
    13, 28, 29, 34, 39, 69, 71, n7, n11, n14
  cultivation 6, 85
MIMI 3
Mimosaceae, see /bírgè/, /teli/, túè/
/mine/ (Eragrostis cilianensis): a wild grass
  26, 71
/míɲà/: the stone of the /kíè/, /míɲà ásán/: a
  nougat, /míɲà bùr/: the kernel of the /kíè/ 20
/mîs/: cat 42
MISALLAKHAT (place name) 53
'miskīn' (Ar.): poor 27
'mogdum' (Ar.): representative of a local chief
  44
money 8, 86
'morah' (Ar.) 77, n75
/mori/: ocarina, clay flute 76
MORTCHA 3
'moxet' (Ar.), see /mádì/
MOURDI 3
/mùgùlí/: the end of the rainy season 33, 68
/múlúkíé/ (Ar. 'muluxiya'; Corchorus olitorius) 26
'muluxiya' (Ar.), see /múlúkíé/
MURRO 27
MUSBAT (vill., wadi and wells) 7, 32, 36, 41,
  44, 53, 54, 68, 75, n63
/musumará/ (Ar. 'talha'; Acacia seyal) 26,
  35, 38, 86
'muxīd' (Ar.), App. Boscia senegalensis

'nabak' or 'nabag' (Ar.), see /kábárà/
'nabag al fīl' (Ar.), App. Ziziphus mucronata
'naga' (Ar.), see /márdì/
'najila' (Ar.), App. Cynodon dactylon
'nāl' (Ar.), App. Chrysopogon sp.
/námár/, /námárdà/ (Ar. 'tumtum' or 'tundub';
  Capparis decidua or sodada) 22, n42
NANA (wells) 55
'nataš' (Ar.): Crotalaria thebaica 36
natron 6, 8, 44, 73, 76, n7
Neurada procumbens, see 'saadan'
'nisā' (Ar.), App. Aristida papposa
/niuw/ (gōz) 53, n64
'niyāda' (Ar.), App. Abutilon
/nógò/ (Ar. 'askanit'; Cenchrus biflorus) 16,
  17, 36, 45, 84, n83
/nógù/ (Ar. 'siget'; Cyperus rotundus) 23
  /nógù bei/ 23
  /nógù keni/ 23, 24
'noram' (Ar.), App. Farsetia ramosissima
/núr/, /núrdà/ (Ar. 'kurmut'; Maerua crassifolia) 22, 35, 44, 85, 86, n21, n42

/nyaRi/ (Ar. 'okra' or 'bāmya'; Hibiscus esculentus) 6, 16, 25, 87
/ɲárì/, /ɲálRà/ (Ar. 'giddem'; Grewia populifolia) 19, n18
/ŋàl/ (Ar. 'timilexe' or 'timileiki'; Gynandropsis gynandra) 17, 23, 26

Ocimum hadiense, see /ósù súlí/
/ohurra/ (clan) 47
oil 23
/ó kèrr/: curdled milk 76, 78
    /ó kiláû/: porridge with millet flour and milk 79
    /ó savi/: powder milk 79
    /ó tògòmó/: whey 79
/ókkò/: a rake 14
'okra'(Ar.), see /nyaRi/
OMDURMAN 8
'omodiya' (Ar.): district ruled by an 'omda' or 'umda' n65
/órì/: waterskin 38
ORORI (wells) 53
/órù/ (Ar. 'battix'; Colocynthis citrullus): water-melon 24, 72
ORU - BA (vill., wadi and wells) 7, 16, 35, 44, 49, 51, 71-72
Oryza breviligulata, see /tómsò/
/ósù súlí/ (Ar. 'fileya'?; Ocimum hadiense) 25

Panicum: wild grass
    P.laetum, see /sábà/
    P.turgidum 36
pastures, see grazing
Pennisetum tiphoideum, see /bíní/
polygyny 10
pools 3, 37, n46, n47, n48
population n1, n51
potters 7
prices n7

'rahad' (Ar.): large pool n46
rainfall 2, 3, 4, 5, 83, n32
    amount of 32-33
rake 14, 17, 23
RAKIB (place name) 53
Ramadan: the ninth month of the Islamic year, a month of fasting 21, n31
RAG - RAG or RAK - RAK (vill.) 70, n48
Rhamnaceae, see /kábárà/, /kíè/
Ricinus communis, see /eigo-biRi/
'rifai': a breed of donkeys 45, n58, n92
rights
    of gathering 17-18
    of herdsmen 17-18
    on wells 45-46
'rotl' (Ar.): a weight n7

'saadan' (Ar.: Neurada procumbens) 36
'saɛada' (Ar.), App. Cyperus nitida

/sábà/ (Ar. 'kreb'; Panicum laetum): a wild grass 15, 34
'sabka': salt pastures 55, 69, 72
sadaga 18
'safil' (Ar.) n64
'sahab' (Ar.), App. Anogeissus schimperi
SAHARA 49, 51
'sakarnaba' (Ar.), App. Cassia mimosoides
'sakia' (Ar.): water-wheel 88
'salaɛlaɛ' (Ar.), App. Cissus quadrangularis
'saleyan' (Ar.): Triraphis pumilio 36
salt 6, 8, 48, n7
Salvadora persica, see /ùí/
Salvadoraceae, see /ùí/
'sanasana' (Ar.), App. Cassia sp.
SANUSI 2
Schmittia pappophoroides 36
Schouwia purpurea n15
Sclerocarya birroea, see /génè/
season 32-33; see also /álgì/, /bógóù/, /dábó/, /gyé/, /ìrsásì/, /tàrbà/
SEDDI (place name) 70
'sɛf' (Ar.), see /álgì/
/seli/: sword 47
SELI - MARA (vill.) 85
SENDI (place name) 54
SERDE - BA (vill.) 7
/seri/, see /kābāš/
/séRí/ (Ar. 'jizu'): a kind of pasture 36, 46, 49, 53, 68, 69, 70, 75, 84, n64, n66, n68
    see also 'jizu'
'seyal' (Ar.), see /kedira/, App. Acacia tortilis
sheep 40, 41, 43, 44, 50, 51, 52, 75, n7
    morphological features 44
    see also flocks and herds, seasonal movements, milk, water
shepherd, see herdsmen
SHIGET KARO or SHIGEIT KARO (place name) 54
'sidr' (Ar.), App. Ziziphus spina-christi : a jujube tree
'siget' (Ar.), see /nógù/
Simarubaceae, see /gíè/
'simēma' (Ar.), App. Aristida pallida
'simsim' (Ar.), App. Sesamum alatum
'simsim al jimal' (Ar.), App. Sesamum indicum
'singid' (Ar.), App. Seddera
SINI - OMU (place name) 69, 70
slaughter-houses 91
soils 34, n34
/sóɲò/, /sóŋóRà/ (Ar. 'baxšem' or 'baxšamay'): a tree 19-20
SUDAN 4, 5, 6, 7, 8, 10, 13, 17, 26, 28, 39-41, 43, 49-51, 68, 72, n1
SUENI (dar) 43
SUNJABAK (wells) 69
/sundu/ n15

117

'sunut' (Ar.), App. Acacia nilotica
SURA - JARI (place name) 70
'sureh', var. 'surei' (Ar.) n21; App. Maerua crassifolia
'šādūf' (Ar.): the beam-well n94
'šao' (Ar.), see /ùí/
'šelīnī' (Ar.), App. Zornia diphylla
/šige/(Ar. 'goz'): sandy soil 34, 51 see also 'goz'
'šii' (Ar.), see 'ediseru'
'šitā' (Ar.), see /dábó/
'šuxiyat' (Ar.), App. Combretum
'šuwal' (Ar.): sack or skin n7

/taga-sao/: skin of camel's neck 79, 80
'talha' (Ar.), see /musumará/
TAMA 3, 27
'tamada' (Ar.), see /kū̃ná/
TAMA - JORA (wadi) 55
'tamalēga' (Ar.), App. Amaranthus graecizans
Tamarindus indica, see /médèr/
tar 17, 19
/tárà̀/ (Ar. 'drese'; Tribulus terrestris) 17, 19
/tărbà/ (Ar. 'darat'): season of harvests 4, 32, 33, 46, 52, 53, 68, 69, 70, 71
TARI - MARA (pool) 52
TARINGEI ROCK, /tríŋè/ 34
TASSILI 88
technology 9
TEDA 18, 27
TEDI - BA, /teɒi-bà/, see Bir Atrun
TEGA (wadi) 52
TEKWOILA (place name) 54
/teli/ (Ar. 'haraz'; Acacia Faidherbia albida) 35, 38, n39
TERI - BA (wells) 69
TIBESTI 41, 89
Tiliaceae, see /gúgúr/, /kòrfú/, /múlúkíé/, /ñáʀì/, /sóŋò/(?)
'timileiki', var. 'timilexe' (Ar.), see /ŋâl/
TINÉ (vill., wadi and wells) 3, 7, 44, 49, 68, 69, 70, 71, 75
'tōb' (Ar.): rectangular piece of cotton cloth 77
/tógórò/ (Ar. 'gafal'; Commiphora africana) 23, 26, 35, 38
/tógóʀíà/: stones inside the drupes of /tógórò/ 23
/tógúì/: cloth 21
/tómsò/ (Ar. 'am-belele'; Oryza breviligulata): wild rice 16, n14
'tomur el abid' (Ar.), see /kòrfù/
/tow/: lending and borrowing 48
trade 7-8, 13, 24, 28
  border trade 8
transfer
  animals n59
  population 93, n101

transhumance 9, 31-32, 49-72, 81-82, 83, 85, 94, n66
Tribulus longipetalus, see 'guttub'
Tribulus terrestris, see /tárà̀/
Triraphis pumilio, see 'saleyan'
tsetse fly 2, 93
TUAREG 27, n16, n78
/tubunu/: pools in the bed of a wadi n46
/túdù/, /túdùrà̀/: Coccinia grandis 24
/túễ/ (Ar. 'kitir'; Acacia senegal) 5, 26, 35, 85, 86, n42
TUER (dar) 33, 37, 43, 49, 50, 51, 54, 70, 90, n73
'tukā' (Ar.), App. Grewia (?)
'tumām' (Ar.), App. Panicum turgidum
'tumtum', var. 'tundub' (Ar.), see /námár/
'tundub' or 'tuntub' (Ar.), see 'tumtum'
TUNDUBAY (vill.) 49, 50, 51, 70, 71, n48
TUNJUR 43, 86, 87, n31
TURBA (wells) 54
TURDA - MYE (place name) 69, 70
/turoŋda/ (clan) n16
/túrù/, /túrdà/ (Ar. 'andarab', 'andrab' or 'ındarab': Cordia Rothii or gharaf) 23, n8
TUWANIS (place name) 53

/ùí/, /ùíʀà/ (Ar. 'šao'; Salvadora persica) 24
UM BEIRI (vill.) n63
UM BURU (vill.) 7, 33, 49, 75, 90, n3
'umda' or 'omda' (Ar.): a petty chief 69, n65
UMDUR (lake) 37, 71, n48
UM HARAZ (vill., wadi and wells) 53, 54, 68
UM KADDADA (vill.) n63
UM MARAHIK (vill.) 7, 53
UM SHEDAR (pool) 37, 52, n47, n63
'umm abaka' (Ar.), App. Setaria verticillata
'umm ɛabaka' (Ar.), App. Euphorbia
'umm ɛag' (Ar.), App. Digitaria
'umm barko' (Ar.), App. Arum sp.
'umm dibēgu' (Ar.), App. Sporobolus festivus
'umm dofūfū' (Ar.), App. Pennisetum
'umm fisēsyāt' (Ar.), App. Fimbristylis
'umm furaw' (Ar.), App. Panicum
'umm hamdī' (Ar.), App. Oscigonum
'umm mušbat' (Ar.), App. Asparagus
'umm rihan' (Ar.), App. Ocimum
'umm tiwēgyāt' or 'tuwēgyāt' (Ar.), App. Cyperus sp., Kyllinga
'umm tuk' (Ar.), App. Cyperus
UÑAY (Zaghawa) n63
'urgasī' (Ar.), App. Chrysophora
'uršāš' (Ar.), see /ìrsàsi/. The var. 'uršāš' is used in Dar-Fur; the form 'rušāš' is more generally used in the Sudan
URU (place name) 71
'ušar' (Ar.), see /kǒrfú/

vegetables 6, 87, n7
vegetation 3, 34-36

Verbenaceae, see /ósù súli/
village 9

WADDAY (sultanate) 2, 41
'warwar' (Ar.), see /dìdì/
watching over
    camels 75-76
    cattle 73-74
    goats 75
    sheep 75
water
    drawing water 38-39, 87-90, n49, n93, n94
    drinking water 76
    storing water 90
    watering camels 76
    watering cattle 74
    watering sheep and goats 75
wells 6, 9, 35, 37-39, 87-90, n49, n93, n94, n95
    dipper 38, 45, 87-89
    rights on 45-46
    see also /bà/, /kúná/

women 13
    female duties 13-14, 27-28, 38-39

'xarīf' (Ar.), see /gyé/
'xašim al ʿama' (Ar.), App. Achyranthes
'xešxāš' (Ar.), App. Stereospermum Kunthianum
'xirwe' (Ar.), see /eigo-biRi/
'xorāb hawša' (Ar.), App. Acanthospermum hispidum
'xušēn' (Ar.): Indigofera arenaria 36

YUMIN ŠIGE (goz) 51

ZALINGEI (town) 82
'zeriba' (Ar.): a thorn enclosure, 73, 86, n75
    see also /guli/
ZIYADIYEH 36, 46, 53, 94
Ziziphus mauritiana, see /kábárà/, /kíè/
Ziziphus spina-christi, see /kábárà/, /kíè/
Zygophyllaceae, see /tárà/